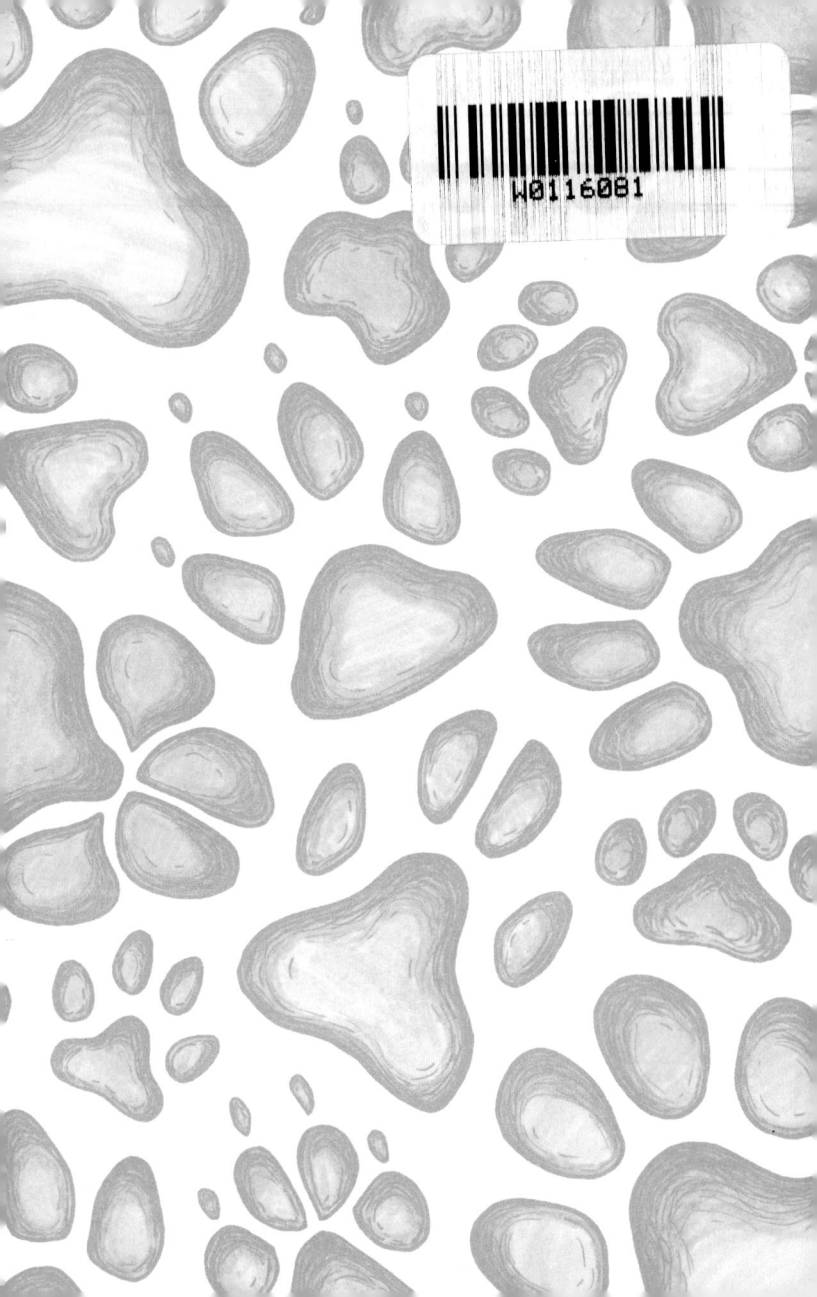

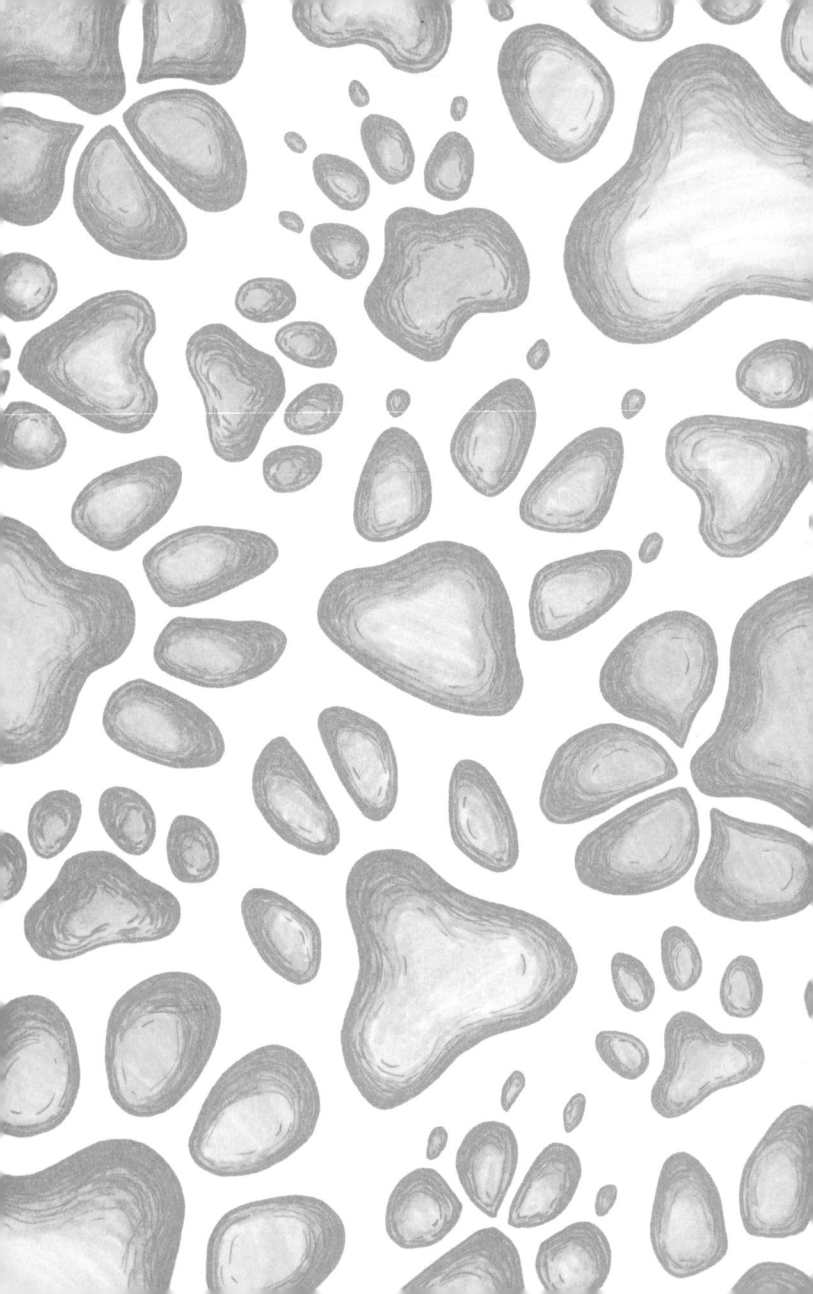

Dogpedia

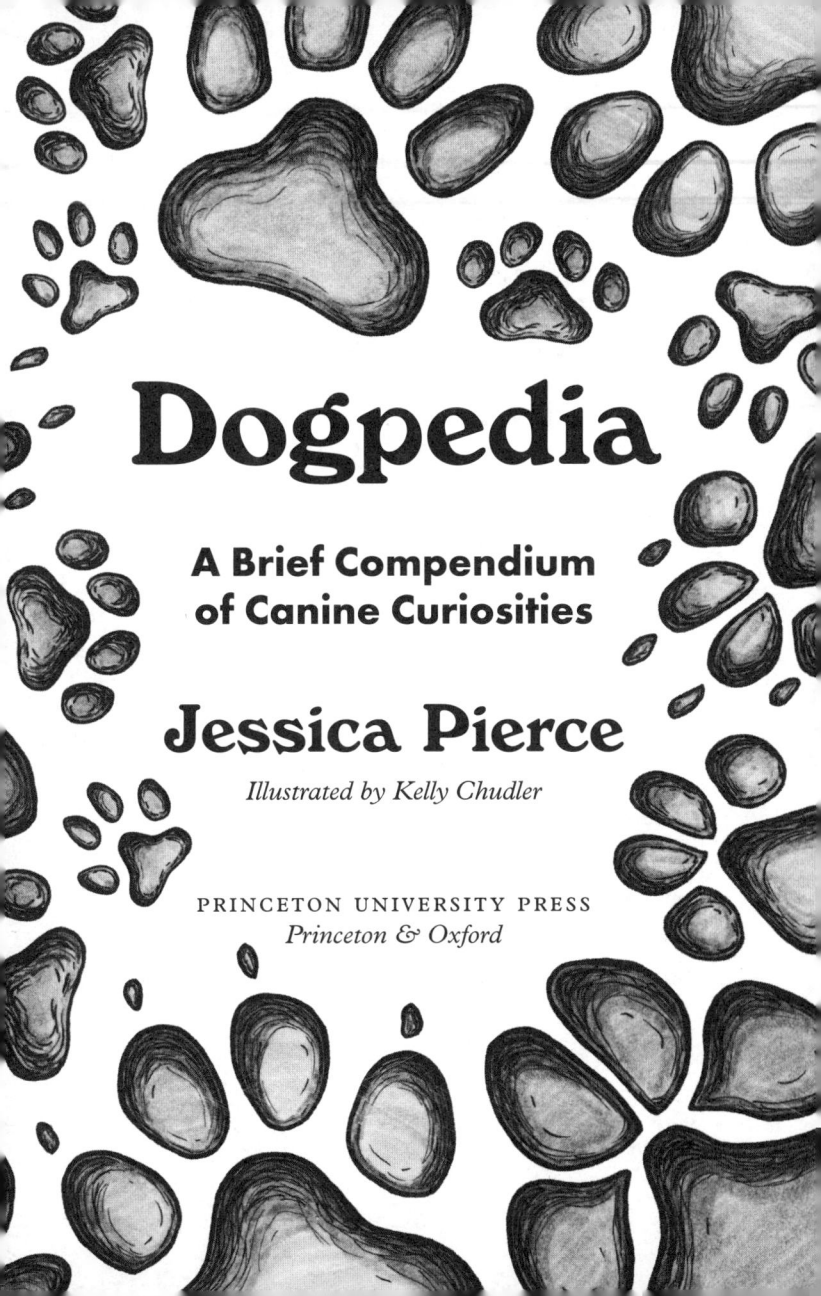

Dogpedia

A Brief Compendium of Canine Curiosities

Jessica Pierce

Illustrated by Kelly Chudler

PRINCETON UNIVERSITY PRESS
Princeton & Oxford

Published by Princeton University Press
41 William Street, Princeton, New Jersey 08540
99 Banbury Road, Oxford OX2 6JX

press.princeton.edu

ISBN 9780691241081
ISBN (e-book) 9780691263274

British Library Cataloging-in-Publication Data is available

Editorial: Alison Kalett and Hallie Schaeffer
Production Editorial: Mark Bellis
Text and Cover Design: Chris Ferrante
Production: Steve Sears
Publicity: Matthew Taylor and Caitlyn Robson
Copyeditor: Annie Gottlieb

Cover, endpaper, and text illustrations by Kelly Chudler

This book has been composed in Plantin, Futura, and Windsor

Printed in China

10 9 8 7 6 5 4 3 2 1

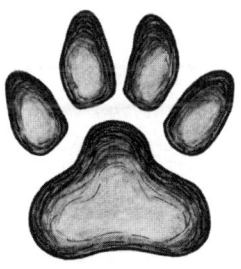

Preface

While I was growing up, my parents sang a lot of folk songs with me and my brother. The one I loved most was about a dog named Blue. Here's how it starts.

Had a dog, and his name was Blue
Bet your life, he's a good dog, too.
"Go on, Blue, I'm comin', too."

According to John and Alan Lomax's *Folk Song U.S.A.*, "Old Blue" is an American folk ballad from the late nineteenth century, likely sung somewhere in the Mississippi Valley. In that time and place, many people would have relied on game animals, including opossum, for survival, and a good hunting dog would have been an invaluable resource. The ballad recounts human and dog hunting together and then sharing the spoils and speaks to the deep friendship forged between human and dog. (A version of "Old Blue" is included in the Lomaxes' *Folk Song U.S.A.*, originally published in 1947.)

Old Blue treed, and I went to see,
There sat a possum in a 'simmon tree.
"Go on, Blue, I'm comin', too."

Blue hunted possum with such zeal and dedication that he grew ill, dying so hard he scratched little holes all around the yard. He was buried with great ceremony. His grave was dug with a silver spade and his face was covered with a possum's face.

I let him down with a golden chain,
Link by link slipped through my hand.
"Go on, Blue, I'm comin', too."

Old Blue is a song of love and loss. I always felt like I knew Blue, and I loved him. My heart broke each time he died. (I admit to feeling a little sad for the possum in the story, too.)

There's only one thing that bothers my mind,
Blue went to heaven, left me behind.
Go on, Blue, I'm comin', too.

And that's how I feel to this day. If there is a heaven, it is most certainly populated by dogs. And this shared human-canine heaven would be consistent with the general theme of human-dog relationships over the course of our coevolution: where one goes, the other follows. Where there are dogs, there are also people. Dogs have led and followed humans to all corners of the planet, and we two species have been essential to each other's survival. We've hunted together and shared our food, and formed a friendship that has lasted for thousands and thousands of years.

Blue touched something in me that felt innate: a sense of kinship with dogs, and a desire to share my life with them. I've done this in the day-to-day sense of living with dogs as companions and housemates, and I've also made dogs a central part of my professional work. *Dogpedia* is my sixth book about dogs, and in many ways, it has been the most interesting and exciting of the books I've written. I thought I knew a lot about dogs, but as I worked on *Dogpedia*, I was continually amazed and delighted by how much I didn't know and still don't know. Dogs are like a magical castle. With each door you open, you find a room full of riches. Beyond the riches you find within each room, you glimpse several new doors. You open these doors and walk through, and more riches appear. You could wander forever.

And wander we have, together with our boon companions. Dogs have played a role in our evolution, history, scientific discovery, medicine, exploration, culture, religious belief and ritual, mythology, music, literature, and art. There is almost no corner of our human past or present that isn't covered in dog hair. In *Dogpedia*, I invite you to wander with me through the world of dogs.

An encyclopedia is a general course of instruction, but the word literally means "training in a circle." This seems apropos, as the entries in *Dogpedia*, although arranged alphabetically, have a way of talking to each other, circling back time and again to each other, often in unexpected ways. My primary goal has been to paint a large canvas of who dogs are as animals, through a selection of entries on dog evolution, biology, ecology, behavior, and sociality. We'll look at why dog noses are so amazing, how dogs manage to convince us to part with half of our sandwich, and how carefully they communicate with each other. We'll examine the diversity of ways in which dogs make a living in the world and how they interact with humans, from being fully ensconced within human homes to living as free-roaming and feral dogs on the far edges of human settlements.

It is impossible to give an account of who dogs are and how they live without also talking about humans. Dogs wouldn't have evolved from wolves if not for their partnership with us, they wouldn't look or behave the way they do without human intervention, and nearly all dogs living on the planet today rely on anthropogenic (human-generated) food resources to survive. By the same token, it is impossible to understand the human species—our evolution, history, culture, and behavior—without also talking about dogs. We would not be who we are if it weren't for our partnership with dogs and without the participation of dogs in various human enterprises. So, although this book is focused on dogs, many entries showcase these codependencies.

A wonderful thing about dogs is that, in Walt Whitman's evocative words, they really do contain multitudes. There is always more to know, there are new ways to see, and there are endless surprises lying in wait beneath all that hair. I hope that every reader, no matter their level of dog expertise, will discover a few new things about dogs and will look at dogs with a renewed sense of curiosity. Although I suspect this book will hold special appeal for people who already love dogs, I particularly invite people who don't know dogs well or aren't too fond of them to read this book, because we inhabit a world shaped by dogs and, love them or not, they are part of our reality and are worthy of our interest.

Before you dive in, here is the final stanza of *Old Blue*:

When I get to heaven, first thing I'll do,
Grab me a horn and blow for Old Blue,
"Come here, Blue, I got here, too."

This book is dedicated to Blue and all the other good dogs out there (and every dog is a good dog).

Abundance and distribution

Dogs are everywhere you look. It may seem like every time you turn around, you see a dog or two. Dogs are one of the most abundant mammals on the planet, with a population numbering close to a billion. They can be found on every continent and in nearly every habitable ecosystem—and even in a few that aren't especially habitable. Where there are dogs there are also, almost without exception, people, suggesting the close interdependence of the two species. People have brought dogs with them to parts of the world where dogs otherwise would not live, including the bitter, frozen edges of the world. Dogs, for their part, have helped humans explore and colonize the far reaches of the planet.

The one billion dogs around the world are not uniformly distributed. Populations of dogs around the world are most highly concentrated in and around urban areas that are also densely populated by humans: New York, Paris, Delhi, Istanbul, Moscow. Yet there are significant variations in dog abundance according to country. For example, the U.S. is a very "dog dense" country, averaging one dog for every 4 people. Saudi Arabia, by contrast, averages only one dog for every 800 people. These huge differences in dog density may reflect the lifestyles and ecological niches of dogs, cultural attitudes of humans toward dogs, the density of human populations, human waste management practices and infrastructure, available food resources for dogs, or some combination of these and other unidentified factors.

See also Conservation impact of dogs; Ecological niches

Activity patterns

One of the ways in which scientists learn about a given species' behavior is by looking at how they use their time. How long a certain kind of animal spends foraging, grooming, and engaging in vigilance behaviors tells researchers about the animal's life history strategies. For a species with whom we arguably have an intimate bond, we know surprisingly little about the activity patterns and time budgets of domestic dogs. When and how much do dogs sleep? When do they hunt and how much time do they spend hunting? How much time do they spend socializing? How does the lunar cycle influence activity? How do the activity patterns of companion dogs differ from those of feral or free-roaming dogs? Working dogs versus nonworking dogs? We really don't have good answers to these questions, in part because the research in this area is relatively sparse and in part because the lifestyles of dogs are so diverse that they are difficult to study.

So, what do we know? Over a 24-hour period, dogs are generally inactive, with bursts of heightened activity. Like other canids, dogs are diurnal, meaning they are generally more active during the day than the night and get most of their sleep at night. Dogs also follow what biologists call a bimodal activity pattern: they have a busy period in the early morning and one during the late afternoon and spend the time in between mainly at rest. Biological factors such as age, sex, and weight seem to influence activity. Younger dogs tend to be more active than older; females more active than males (with some exceptions—in one study, male sled dogs were more active at night than females); and lighter dogs more active than heavier dogs. Environmental conditions also influence activity. Higher temperatures generally decrease levels of activity, as does solitary housing. Perhaps the most consistent finding from the available research is that the activity patterns of domestic dogs are driven largely by their interactions with humans.

A 2021 study offers a window into how scientists study activity patterns in dogs and what kinds of details scientists examine. This study compared activity patterns in several different populations of domestic dogs: owned free-roaming dogs in Guatemala and Indonesia, farm dogs in Switzerland, and family dogs in Switzerland. To track activity patterns, researchers fitted dogs with a collar that measured acceleration, proportion of time at rest, and proportion of time in "moderate" and "high" activity, over a 24-hour period. Data gathered from the collars revealed that although family dogs were far more constrained in their movements, their activity patterns were in the same general range as those of the free-roaming dogs, suggesting that human guardians adapt their own activities to align with the natural patterns of the dogs—or, alternatively, that humans and dogs have similar patterns that can easily be co-adapted. Dogs sleeping indoors or in fenced yards showed greater periods of rest during the night than free-roaming dogs, perhaps because they feel less need to remain vigilant. Interestingly, neutered dogs spent less time in "high" activity than intact dogs.

See also Ecological niches; Free-roaming; Pet

Affiliative behavior

What draws humans to dogs, perhaps more than anything else, is the easy flow of friendly feelings back and forth across the species divide. We generally trust each other and get along. This easy flow is made possible by the continual exchange of affiliative behaviors. Dogs and humans both follow the Golden Rule of sociality: Be nice.

Affiliative behaviors are defined by ethologists as friendly and peaceful acts exchanged among individual animals. In dogs, as in other social mammals, most social interactions (approximately 80 percent) are affiliative rather than agonistic. Some affiliative behaviors of dogs include grooming each other, being in close physical proximity to each other, approaching each other, touching each other, sharing food, and playing together. Dogs also engage in affiliative behaviors toward humans, when they rub against our legs, sit next to us on the couch, lick our faces, or approach us in the park to say hello. Humans, for their part, engage in affiliative behaviors toward dogs, petting them, staying close to them, sharing our snacks. Affiliative behavior provides the glue for social bonding.

See also Aggression; Licking

Aggression

Aggression is defined as hostile behavior directed by one individual at another. It comes into play when animals are establishing and protecting their territory, finding mates, defending themselves, and protecting their children and other family members. Perhaps counter-intuitively, aggression functions to maintain peaceful relations and *reduces* the potential for fights; aggression is not primarily aimed at inflicting harm but avoiding it. For example, the establishment, marking, and defending of territories helps animals know what belongs to whom without having to fight about it every time they move around within an ecosystem. Aggression is part of the behavioral repertoire of nearly all animals.

In social species such as dogs, aggression is often ritualized, as for example in greeting rituals, which involve aggression-*limiting* communication. If two dogs meet, and the more dominant dog presents with a stiff body and confident stance, and the more submissive dog with a low, crouched body, they both know where they stand in relation to each other socially, and have no need to stress about it or spend precious energy trying to figure it out. This is why clear communication is so important between and among dogs. Even actual dog fights, despite the flying fur and explosive energy of the conflict, rarely inflict serious injury—they truly are more bark than bite.

Naturally, humans are especially keen to understand why dogs, especially pet dogs who live in our homes and share our beds, sometimes behave aggressively toward people. It is rather scary to share close quarters with a large, sharp-toothed animal who, not that long ago, was a wolf. Genetic mechanisms are thought to influence various aspects of aggression, such as how quickly a dog will react to a stimulus and what at what level of intensity a stimulus will push a dog to the "snapping point." But exactly how genetics influences aggression is still unclear. During domestication, various components of aggression have been under selective pressure—but in highly confusing and conflicting ways. Dogs have been bred, for example, for low levels of aggression and high latency (the time between exposure to stimulus and response); and dogs have also been bred for high levels of aggression and low latency. (People always want to compare dogs with wolves. Yet such comparisons are tricky, and it is inaccurate to say that dogs are less aggressive than wolves or more aggressive than wolves; dogs are *differently* aggressive, and there is tremendous variation in how dogs express aggression.)

Researchers are trying to untangle the various factors that can lead to aggressive behaviors in pet dogs who live in our homes, including health status (especially pain), age, sex, and past experiences. None of these factors *cause* aggression; but they can play a role in it. Pain and other forms of physical discomfort can increase the expression of aggressive behavior. Older dogs are more likely to behave aggressively than younger dogs. And male dogs show more aggression, on average, than female dogs. The belief that neutering dogs will alleviate or prevent problematic aggression is outdated and not well supported with data. Some studies have shown that neutered males are more likely to be aggressive; some have shown that they are less likely to be aggressive; some have identified no differences between neutered and unneutered, in terms of probability of aggressive behavior. Prophylactic neutering of all young male puppies to "prevent aggression" is not an evidence-based practice; how hormones facilitate behavior is exceedingly complex and not something we currently understand very well. Early weaning is also thought to increase aggressive behavior, as does living in a single-dog household.

Finally, aggression can be shaped by deliberate human interactions. Dogs can be taught by humans to exhibit aggressive behaviors in response to patterns of events, as is seen in trained military and police dogs. The use of punishment-based training has been shown to increase a dog's likelihood of behaving aggressively toward humans, as does repeated cruel treatment by humans.

Within a given context, canine aggression can be appropriate or inappropriate, behaviorally speaking, and both kinds of aggression can be a source of trouble. But it is important to bear in mind the distinction. Sometimes the aggression that leads to a bite is appropriate—a dog felt threatened, the human didn't read behavioral signs, didn't back off, the dog had no choice. "Inappropriate" aggression, or aggression out of context, is a sign that a dog is struggling to adapt to his or her environment. It is not a sign that the dog is mean or vicious or criminal—although we label dogs who behave aggressively in all these ways, and more. These words place an emotional valence on aggression that is likely absent in dogs themselves and clouds our ability to respond objectively.

See also Affiliative behavior; Dominance; Small Dog Syndrome; Submissive behavior

Allelomimetic behavior

I bark. You bark. Everybody bark barks. A single barking dog can soon crescendo into a chorus of howling and yowling spread across a neighborhood. Contagious barking of this sort is an example of *allelomimetic* behavior, in which the performance of a behavior by an individual animal increases the probability of that behavior being performed by other individuals nearby. Humans do this a lot. And so do dogs.

Various canine activities can be allelomimetic: barking in chorus, walking or running together, sleeping together, lying down, getting up at the same time when hearing a suspicious noise (or just because), howling when a siren goes by. This synchronizing of behavior aids in social bonding and helps predatory species like dogs to work in unison, as a team.

Dogs and humans engage in allelomimetic behaviors with each other. If I get up from my chair, my dog will also get up. If your dog lifts his head and looks alert while you are lying together on the couch watching Netflix, you might also become alert, especially if you have been emotionally primed by watching a slasher movie. If your dog then starts growling, the hair might stand up on the back of both your necks.

Anal glands

Anal glands are the bane of existence for many people who live with dogs. The smell associated with a recently "expressed" (emptied) anal gland is like nothing you've ever experienced—pungent, awful, and slightly fishy, according to some reports. But anal glands, or sacs, shouldn't be disparaged; they are very important for dogs, both physiologically and behaviorally. If you looked inside a dog, you would find two small oval-shaped glands, one on each side of the anus, about the size of an almond or grape, depending on the size of the dog. Each gland has a small tube through which brownish fluid is squeezed out when stool passes through the anus and puts pressure on the gland. Each dog's anal gland scent is unique, like a stinky pheromone fingerprint. It may be that the expression of a small bouquet of anal gland scent marks territory or communicates other information—nobody knows for sure. Butt sniffing involves an investigation of the anal glands, among other things. Some scientists speculate that the chemical odors found in the anal sac are different from those in urine, feces, or vaginal secretions and that each smelly substance may play a specific role in communication.

Anal glands are found in many carnivore species, including beavers, wolves, and bears. Wild animals can express their glands voluntarily, to scent mark or as a defensive, skunk-like "Don't mess with me!" display. Dogs appear to have lost the ability to express at will.

However, they sometimes express involuntarily, when startled or extremely anxious.

Pet dogs sometimes suffer from uncomfortable impaction or infection of the anal glands. This can lead to the dreaded and rather ridiculous looking maneuver dog guardians call "butt scraping," whereby a dog skootches forward in a seated posture, desperately trying to relieve itching around the anus. Sometimes the anal glands need to be manually "expressed," a euphemism if ever there was one, and sometimes a course of antibiotics is needed. Anal glands can get inflamed or impacted for a variety of reasons, including obesity, food allergies, chronic skin dermatitis, inadequate diet (not enough fiber), and genetics (anal gland problems are more common in small dogs). Do free-roaming dogs have itchy inflamed butts? Hard to say, but the problem is likely far less common than among pet dogs.

See also Ground scratching; Olfaction; Sniffing

Appeasement

Appeasement signals are ritualized social signals used by animals to communicate their intention to behave in a friendly way. Appeasement signals also allow an animal to communicate submission, to say "Hey, I'm not a threat, don't worry," and thus reduce the potential for injury from a dominant individual. Some appeasement signals include rolling over onto the back, lip licking, and avoiding eye contact. Norwegian dog trainer Turid Rugaas came up with the sticky term "calming signals" to refer to behaviors that dogs use to alleviate tension in their interactions with humans or other dogs, or to self-calm. Rugaas identified about 30 calming signals,

including yawning, nose-licking, averting the head or turning the gaze, walking in a curve, tail wagging, urinating, doing a play bow, and holding ears back toward the neck.

Dogs are socially intelligent, and instead of fighting are much more likely to engage in conflict resolution and de-escalation. Sometimes one dog will act as a mediator between two or more dogs who seem on their way to a fight. The use of calming signals is not unique to dogs. Humans have a repertoire of such behaviors, too, such as yawning or averting our eyes.

See also Submissive behavior; Yawning

Argos

One of the most touching moments in Homer's epic poem *The Odyssey* is the brief encounter between Odysseus and his dog Argos, found in lines 290–327 of book 17. After fighting in Troy and then facing countless obstacles, Odysseus finally returns to Ithaca, disguised in rags so that he doesn't alert Penelope's suitors. As he approaches his home, Odysseus encounters his dog Argos, lying neglected on a heap of manure, covered in fleas. On seeing Odysseus, Argos drops his ears and wags his tail. He, alone, recognizes Odysseus under the beggar's clothes. But Odysseus ignores Argos's greeting, for fear of giving himself away. After Odysseus enters his home, Argos "passed into the darkness of death, now that he had fulfilled his destiny of faith and seen his master once more after twenty years." The scene with Argos is affecting in large part because of the dog's deep love for his human. It is also heart-stabbing because of Odysseus's refusal to acknowledge his canine companion. This passage has made Argos almost synonymous with dogs' unwavering loyalty to humanity.

See also Literature; Loyalty

Artificial selection

Natural selection, which Charles Darwin hypothesized to be the mechanism driving evolution, refers to the process by which organisms best adapted to their environment survive and pass on their genes to offspring. Artificial selection, in contrast, describes the intervention of humans into the selection process to produce genetic change, usually with the goal of enhancing or eliminating certain traits. In the simplest terms, humans pick which animals get to reproduce and which don't, and thus control which traits are passed on to future generations. There is a tendency to think that dogs have evolved purely through artificial selection, while wolves, foxes, and other wild animals have evolved purely through natural selection. But dogs have been shaped by both natural and artificial selection pressures. Artificial selection is sometimes also referred to as "selective breeding."

See also Domestication; Selective breeding

Attachment

Attachment is a social bond between two individuals, based on emotional dependency, and enduring over time. The concept of "attachment behavior" was originally developed within the field of human psychology, to describe a range of behaviors performed by infants to maintain proximity to their mother. Some elements of attachment behavior include safe-haven effects (the caregiver provides a safe haven; the infant seeks to maintain contact or close proximity to the attachment figure), secure base effects (if the infant has a secure attachment to her caregiver, she will be willing to engage in play and exploratory behavior when near the attachment figure). Some scientists believe that dogs can form attachment bonds to humans that are analogous to child-parent bonds, and that this capacity for attachment to humans is unique to dogs as a species and is linked to domestication.

Avoidance

An avoidance behavior is an attempt to withdraw from an aversive stimulus, whether by actively fleeing, freezing in place, or trying to hide. Avoidance is one of the most important categories of dog behavior for humans who keep dogs as pets, and one of the most often overlooked. A leashed dog, for example, may signal that she is uncomfortable meeting another dog by looking in the other direction and moving in an arc to maintain distance from the other dog. These signals may go unnoticed by the human at the other end of the leash or might even get interpreted as the dog lollygagging instead of paying attention to "leash manners" (defined by human trainers as "walking in a straight line at the exact same pace as your human guardian and at the exact right distance from her heels").

Balto

A Siberian husky named Balto is famous for having led a team of sled dogs on the final leg of a run which brought a shipment of diphtheria antitoxin serum from Anchorage to the town of Nome, Alaska,

averting a potentially catastrophic epidemic in the small town. The 1925 serum run was the inspiration for the Iditarod, the most famous sled dog race in the world. The grueling race covers about 1,000 miles of Alaskan wilderness in the dead of winter. Balto is also a furry reminder that the nineteenth-century gold rushes in Alaska and Canada would not have been possible without the help of sled dogs. Indeed, sled dogs have been exceedingly important in many human endeavors, including just plain old survival.

See also Landrace; Working; Zhokhov Island

Barking

It would be an overstatement to say that barking is unique to dogs. Other canids certainly bark. But barking is of particular importance to dogs, and dogs are the only canid (as far as we know) that uses barking specifically to communicate with humans. Dogs use acoustic signals, including the bark, to "speak" directly to humans: research has found that stray and feral dogs vocalize less than dogs living in human homes as pets. Dogs also use barking to communicate with each other. Barking serves a range of purposes in various contexts: defending an area, warning, greeting, playing, and drawing visual attention to the barker.

In the most general terms, barks are short, explosive signals, often coming in quick sequence. But the acoustic structure of barks varies from one context to another. For example, rapid, mid-range barks are used to signal an alert, stutter barks often initiate play, and a low-pitched, slow bark might be used in response to a threat. Barks can be highly variable, differing between breeds and individual dogs. Many dog guardians can identify a range of different barks: Rufus the dog, for example, may have a special "play with me" bark, a "there's another dog out there" bark, a "FedEx Guy!" bark, and an "I can't believe you left me at home!" bark.

Barking has been under selective pressure during domestication. Some breeds of dog have been bred to bark a lot, for example in the context of guarding (German shepherd). Other breeds bark relatively infrequently (shar-pei), or not at all (the basenji), although these "quiet" breeds still have an extensive repertoire of acoustic signals. Baying, a unique sound made by some breeds of dog, is a long, deep, and almost haunting cry, such as bloodhounds make when tracking.

Although barking is a natural part of dogs' behavioral repertoire, it is a behavior that humans find difficult to countenance. Normal levels of barking are often labeled a behavioral problem by guardians, even though barking is something all dogs need and want to do. How much barking is judged to be "too much" is quite subjective. Nevertheless, there is a point at which barking dips from normal into excessive or obsessive, which can be a sign that a dog is suffering from psychological distress. Barking is also a significant flashpoint between dog-keeping and non-dog-keeping humans. Barking dogs have been known to send people into a state of homicidal mania.

See also Allelomimetic behavior; Communication

Begging

Dogs are extremely skilled at asking for—and getting—what they want from us. When dogs ask for stuff, we often label it "begging." They are especially good at asking us to share food and asking us to give them some love. Begging can involve standing close to us, looking up at us with big, sad eyes, whining, scratching or pawing our leg, drooling, and barking. Dogs have special facial muscles that help them beg from humans—the so-called puppy-dog eye muscles—which make their eyes look bigger, more paedomorphic (childlike), more needy and appealing than usual. Dogs also lock eyes with us, which can be unsettling and will often result in us sacrificing a chunk of our sandwich. Begging is a normal canine behavior—puppies beg for food and attention from their mothers. But it is also a learned behavior; every time we give in to a begging dog, we reinforce for them that begging works. So, they'll try it again. All dogs (pet, village, street, free-roaming) who solicit food directly from humans engage in some degree of begging.

In the context of pet dogs, begging is almost uniformly labeled an unwanted, even a "bad," behavior. The number one piece of advice given to dog guardians with a begging dog is to ignore the behavior. Oh, but that's hard to do.

See also Domestication; Puppy-dog eyes; Solicitation behavior

Boji

Boji is one of our most colorful contemporary dog celebrities and one of the few street dogs to achieve global notoriety. He is famous not for starring in films or accumulating a huge social media following (although he has done the latter). Boji rose to fame for being plucky, independent, and charming, and for his habit of regularly riding around on the public transport system in Istanbul, Turkey. Every day, Boji would commute around the city, climbing onto and off ferries, subways, trains, and trams, seeming to know exactly where he was going and how to get there. A photographer who followed Boji for a day reported that he passed through no fewer than 29 metro stations. Boji was always polite, poised, self-confident, and friendly. He was, in other words, a perfect citizen. He was adopted in 2022. His human guardian has insisted, in interviews, that Boji still has the freedom to roam.

See also Hayırsız Ada Dog Massacre

Breed

"What kind of dog is he?" This is a common conversation opener in countries where large numbers of dogs are kept as pets. The interlocutor doesn't typically mean "Is your dog the kind of dog who is suspiciously likely to nip the hand of unfamiliar people?" or "Is your dog the kind of dog who likes to roll in goose poop?" What is generally understood by this question is, "What *breed* of dog is he?" Aside from the possible offense to dogs of mixed parentage, the question is innocent enough. Yet it reveals a cultural obsession with the notion of dog breed that has important consequences for dog welfare.

A "breed" is generally understood to be a group of domesticated animals or plants that have shared behaviors or physical traits that distinguish them from other groups of the same animal or plant. Within the context of modern pet-keeping practices, a breed is a group of dogs with identifiable, heritable external characteristics such as body shape, fur color, and length of muzzle, as well as behavioral characteristics such as a proclivity for retrieving objects, which can often be traced to the functional role of ancestral source populations. There are now roughly 1,000 different breeds of dog. Fewer than half of these are officially recognized by breeding clubs.

Although many prospective dog owners use breed temperament and behavioral descriptions generated by kennel clubs or find-your-perfect-doggie-match apps to decide which breed of dog best fits their lifestyle, detailed scientific analysis of behavioral characteristics suggests that behavior can be quite variable within any breed. According to one study, how well dogs respond to human direction ("biddability") was the most heritable trait, by breed. (Basset hounds got last place on "biddability"; Belgian Malinois won out as most "biddable.") But even this trait varied significantly from one individual dog to another. The take-home message here is that breed stereotyping—"The XX is an affectionate family dog, who loves nothing more than to curl up next to you on the couch"—is nonsense.

See also Landrace; Mixed-breeds; Pedigree; Purebred

Brown Dog Affair
The Brown Dog Affair is often cited as the beginning of the anti-vivisection movement. Given the affectionate relationships humans have always shared with dogs, it is perhaps not surprising that the use of dogs in painful scientific experiments would galvanize a response.

Before *anti*-vivisection, there was vivisection. Let's situate ourselves in the 1800s, in Victorian England. Huge strides were being made in the study of anatomy and physiology; there was also a newfound interest in using live animals as experimental subjects and for teaching demonstrations. This was certainly not the first time in history that people had done horrible things to dogs in the name of science and medicine. But something shifted in this historical moment. The use of live dogs and other animals became institutionalized; vivisection *became part of the scientific method.* Vivisection is the dissection of live animals (Latin *vivus*, "alive" + *secare*, "to cut"). "Vivisection" entered the English vocabulary in the early eighteenth century. ("Dissection," with the meaning of cutting open a nonliving animal or a plant to examine tissues or organs, is from ca. 1600.)

In response to the burgeoning practice of medical researchers collecting stray dogs off the streets and vivisecting them, activists fought back, seeking to protect dogs from suffering by, at the very least, convincing researchers to apply anesthesia before surgeries. The National Anti-Vivisection Society was established in London in 1875 by Frances Power Cobbe, as part of a larger anti-cruelty movement associated with Henry Bergh, the founder of the American Society for the Prevention of Cruelty to Animals. As a historical sidenote: It is interesting that the anti-vivisection and animal sheltering movements arose together, emerging from a shared historical, cultural, and social soil—the same fertile ground out of which the suffrage and abolitionist movements grew.

The Cruelty to Animals Act, pushed through the British legislature by Cobbe and others, created a licensing system for people using live animals in experiments. It required that animals be anesthetized during experiments and killed immediately afterward. Nobody really followed the rules, though. One particularly blatant violation was research conducted by Ernest Starling and William Bayliss. They were following up on work done by Ivan Pavlov on pancreatic secretions and how these were controlled by the nervous system. Bayliss and Starling, in front of an audience of students, which included two women activists disguised as men, cut open the abdomen of a small brown terrier. The dog already had a wound from a previous experiment (against the rules), and was, according to witnesses, clearly not anesthetized (also against the rules): the dog was howling in pain during the entire event (not against the rules, per se, but horrifying nonetheless). The two women wrote a book exposing cruelty in science, including the horrific brown dog experiment. A lawsuit was filed against Bayliss and Starling. Although it was unsuccessful, the lawsuit created a publicity buzz which galvanized the antivivisection movement. The experiment and its fallout became known as the Brown Dog Affair of 1907. Sadly, more than a century later, vivisection continues to be routinely practiced on dogs.

See also Saliva

Cartoon characters

What percentage of famous animal cartoon characters have been dogs? Unfortunately, this burning question remains unresearched, so we are left with rampant speculation. Dogs do seem to be heavily overrepresented in cartoons, relative to other animals, perhaps a signpost of dogs' cultural importance. Depending on your generation, you may be familiar with Scooby-Doo, Snoopy, Goofy, Astro, Mr. Peabody, Underdog, Marmaduke, Clifford the Big Red Dog, Brian, Muttley, Pluto, Santa's Little Helper, Ren, Gromit—the list goes on and on. Most of these cartoon dogs are goofy, gullible, and liable to waltz headlong into embarrassment and danger. A rare few (Mr. Peabody and Snoopy) conduct themselves with aplomb. And many are part of a human-dog partnership or are members of a human family (Clifford, Marmaduke, Scooby-Doo). To date, there have been no popular cartoons featuring free-roaming or feral dogs, a significant oversight. The closest we have is a stray dog named Tramp, from Disney's animated movie *Lady and the Tramp,* who, despite his street-smarts and bravado, secretly longs to become a pet.

How many famous cartoon cats can you think of? A few come to mind: Felix the Cat, Garfield, Oliver, Tom, Sylvester, Tigger. But they simply don't roll off the mind with the same ease as dogs, do they? How many famous fictional guinea pigs can you come up with? (Hint: Norman, G.P., Darwin, Juarez, Blaster, and Hurley.)

See also Movie stars

Cerberus

Dog-bodied or dog-headed deities spend a lot of time guarding the gates of the Underworld. One of the most well-known of these creatures is Cerberus, who in Greek mythology is a multi-headed dog who stands guard outside Hades, lest any dead people try to escape (or live people try to sneak in). Cerberus, or the Hound of Hades, is usually depicted with three heads; Hesiod, however, gave Cerberus 50 heads. Often called Hell-hounds, these mythological hounds take on two opposing roles: either they guard the gates of hell and are a servant of the devil, or they work for god(s) and serve as guides to help people's spirits get safely through the underworld and to their final destination in a happier place. Hellhounds are often depicted as black, sometimes with red eyes, and are overgrown and monstrous.

Hellhounds are symbolic of many things, the most obvious of which is a deep human ambivalence toward dogs: they are sometimes feared as the devil and sometimes worshipped as gods. In Chinese mythology, a dog is said to have given the first provision of grain; the dog is one of the twelve totem creatures for which years are named. (The next Year of the Dog will be 2030.) In English and Nordic folklore, the guardian spirit that oversees and protects a churchyard, called a Church Grim, takes the form of a black dog. The Egyptian god Anubis was often depicted with a jackal's head and led people from their tomb through the desert and into the land of the dead.

Hellhounds and other scary, supernatural black dogs appear all over the place in literature, film, television, and computer games. In Goethe's *Faust*, the devil Mephistopheles first appears in the form of a black poodle, a curious choice. One of the most famous of Arthur Conan Doyle's mysteries, *The Hound of the Baskervilles*, revolves around an ancient family curse and a spectral, demonic hound. Hellhounds make an appearance in various spooky movies, including *The Omen*, in which Damien the Antichrist is served by a black rottweiler.

See also Cynanthropy; Literature

Chewing

A canine behavior that has garnered a great deal of attention, mainly because of the propensity of pet dogs to apply the behavior to items we have not labeled food and which we value for sentimental or financial reasons, is chewing.

Chewing is something that all canids do, as part of their feeding behavior: they chew on bones to get at the marrow, they gnaw at tendons. Beyond its obvious function in helping dogs consume prey animals, chewing also likely functions to help keep dogs' teeth clean and may be inherently satisfying. Dogs kept as pets still have a behavioral need to chew, even if their food comes in the form of easily crunched-up kibble. Without an animal carcass at the ready, they will find other chewing outlets. Chewing may be a form of stress relief for pet dogs, or a way to deal with feelings of frustration or boredom; excessive or obsessive chewing can be a sign of compromised welfare. Puppies are notorious chewers, and there's a good explanation for this: teething hurts and chewing helps relieve the pain. Chewing is a behavior that can be difficult for humans to direct in ways that don't annoy us. Thus, the enormous and highly profitable industry dedicated to addressing this behavioral need: bones (real and fake), rawhides, antlers, pig ears, bull penises (which are called "pizzles" so nobody freaks out), dried chicken feet, Kongs™, and so forth. Professional opinions about which of these items are appropriate and which dangerous are all over the board.

See also Diet; Emotions; Kibble

Clicker training

One of the most iconic instruments of dog training is the so-called clicker. The clicker connects dog and human through a piece of handheld technology about the size of an eraser: encased in a small plastic box is a metal strip that makes a distinct two-tone clicking sound when pressed. The clicker is used to cue a dog that a certain behavior is what the human wants; the click is typically followed by a food treat. The point of the device is to smooth the potentially clumsy timing of human communication with a dog, not leaving a dog to guess which behavior a given treat is rewarding. The click says, "Yes! That's right!"

Clickers are generally associated with rewards-based training philosophy and are an example of reinforcement training, based in the broader framework of B. F. Skinner's theory of operant conditioning. Without going into painful detail on learning theory and behaviorism, the clicker is what Skinner would have called a "bridging stimulus": the click is used to bridge a delay between a conditioned stimulus and an unconditioned stimulus. Naturally, there is controversy over what exactly a bridging stimulus is, whether clickers are bridging stimuli or marking stimuli, whether they improve learning, and how we should measure learning (e.g., speed of learning acquisition vs. whether the learning lasts over time). Basically, we've flown into a hornet's nest of differing and contradictory opinions about how best to train dogs to respond to human cues.

The clicker is one of many remote training devices, and is decidedly low-tech. A higher-tech and highly contentious tool for remote training is the shock collar, which uses electronic stimulation of the tissue of a dog's neck as a stimulus and is associated with punishment-based or aversive training, or what is now sometimes euphemistically referred to as "balanced training."

See also Collars; Pet

Collars

As long as humans have lived closely with dogs—and especially as humans have sought to keep dogs captive as pets or workers—dogs have been fitted with collars. The primary function of the collar has always been pragmatic, but collars have also long been an expression of human artistic creativity, an adornment that expresses both our affection and our sense of style, and perhaps also our social status. Collars and leashes may have been developed by the Sumerians (roughly 3300 BCE)—or at least this is where some of the earliest depictions of dog collars and leashes have appeared. Egyptians created artistic collars of pure gold, as well as practical collars, such as the slip collar for racing and warring. Thick studded collars were thought to protect dogs from attacks by wolves. A gypsum wall panel relief from 645–635 BCE held by the British Museum in London—"Assyrian Huntsmen with Hounds"—shows dogs with collars, attached to leashes held by the huntsmen.

Although not a new contrivance by any means, collars are more ubiquitous now than ever. Nearly all pet dogs—and even some free-roaming dogs—wear a band around their neck. A collar signals a dog's status as "owned," not "stray." We now have a zillion different kinds of collars: flat, prong, slip, choke, martingale, shock, e-stim, GPS, and "smart," each with a unique set of welfare problems for dogs. Research by veterinary scientists has shown that all collars—even a simple nylon flat collar with a clasp—can damage the delicate structures of a dog's neck. Harnesses are generally considered a far safer choice for dogs being walked on a leash.

See also Leashes; Pet; Stray

Communication

In simple terms, communication occurs when an animal does something that makes another animal change his or her behavior. In most communication, there is a sender, a receiver, and a signal. Often both the sender and receiver are dogs. Other times, the sender is a dog, and the receiver is a person. And sometimes the sender is a person, and the receiver is a dog. (There are, of course, many other iterations as well. Dogs certainly communicate with cats, sheep, horses, foxes, squirrels, and other species of animal with whom they interact.) A few examples of signals that might be sent by dogs to each other are chemicals left in a spot of urine, a facial expression such as a lifted lip, a noise such as a growl, a body position such as a play bow, and the touch of one dog licking the mouth of another. Communicative signals are often described as discrete events ("ears erect"), but dogs often send what are called composite signals, which combine several different modalities (e.g., visual and chemical, such as when a dog scratches the ground after pooping).

Dogs send visual signals by modifying different parts of their body, such as hair, ears, tail, posture. They can make themselves look bigger by standing erect or by raising their hackles and can make themselves look smaller and less threatening by pulling their ears back and lowering their body. Tails can be held high to indicate confidence or arousal, can be loosely wagging to indicate friendliness, or can be low wagging to signal nervousness or motivational conflict.

A lot of information about mood and intention can be communicated by a sound or combination of different sounds. The signal with which we are perhaps most familiar is the bark, which can communicate a whole range of information. A bark may be used to get another dog's attention, to invite play, to communicate a threat. A bark also communicates a dog's size, can be saying something like "Don't mess with me—I'm bigger than you." This can be beneficial to dogs on the receiving end of the communication because it can help them judge from a distance or without visual access whether they would be at a disadvantage in an agonistic encounter. Vocal communications are extremely nuanced, varying in pitch, intonation, volume, prosody (the "musical" aspect of the vocal signal, which includes timing, frequency spectrum, and amplitude). Dogs may use acoustic signals other than vocalizations. Ground scratching, for example, is typically described as a visual and olfactory signal, but it may have an acoustic component.

Dogs also communicate with what are called contact behaviors. Tactile communication is used in agonistic encounters, such as when a dog places one of his legs or his body on the back of another dog. Touch is also extremely important in affiliative encounters, as a way of building social cohesion. Dogs position their bodies close together or even touching during rest, they nuzzle each other, lick each other's faces, or groom each other. "Nibble" grooming involves rapid open-close jaw movements and just the front teeth, and almost looks like a dog's teeth are chattering from cold. Dogs nibble groom each other and sometimes themselves. Skin has complex receptors, and dogs have touch receptors at the base of every hair. One of the most important touch receptors for dogs are the vibrissae, more commonly referred to as whiskers.

Chemical communication can be direct, such as two dogs sniffing each other's butts as they meet. It can also be indirect. Odors are a way of communicating without being physically present: a dog can leave a pee message to be "read" later, by passersby. Dogs not only leave chemical signals with urine, but also with feces and glandular secretions (e.g., from the anal sacs).

Do dogs use the same signals for communicating with humans as they do for communicating with each other? To some extent, yes. But because of their close and long-term association with humans, dogs have also developed specialized skills for relaying information to us and for reading our cues. For example, some scientists believe that eye contact in dog-human communication is unique. Whereas dogs might avoid eye contact with each other, to avoid communicating a threat, dogs often gaze fixedly at humans—especially those they like—and will use the direction of human gaze as a source of information ("the human is looking over there; maybe that's where the treat is hidden"). Other research has shown that human gestural cues, such as pointing a finger or using a hand signal for a command, may have particular salience. In one study, dogs were given contradictory cues, one verbal command and one gestural. Dogs relied more heavily on the gestural cues.

When it comes to pet dogs, people often wonder whether communication skills vary by breed. And in fact, they may. Certain traits which humans have selected for aesthetic reasons may have the unintended consequence of reducing communicative nuance. *Brachycephalic* breeds, for instance (from Greek *brakhys*, "short" + *kephalē*, "head")—dogs with shortened skull bones that give the nose and face a distinctive "smushed-in" appearance, such as pugs and French bulldogs—have less flexibility and range in facial expression than dogs with a more wolf-like skull and muzzle, and as a result are likely handicapped in communications involving wrinkling of a nose or raising of eyebrows. Likewise, dogs with very short tails may not be able to communicate mood and intention through tail posture as clearly as their long-tailed kin.

See also Allelomimetic behavior; Anal glands; Barking; Ears; Facial expressions; Gazing; Ground scratching; Hackles; Puppy-dog eyes; Olfaction; Scent marking; Tails; Urine; Wagging; Whiskers

Companion animal

"Companion animal" began appearing in the literature in the 1960s, and since then has gained increasing popularity as the favored term to refer to dogs who live within human homes and in close companionable relationships with humans. Often used in concert with "guardian" as alternatives to "pet/owner."

See also Pet; Owner

Conservation impact of dogs

A key text on the conservation impacts of dogs on wildlife, by Matthew Gompper, opens with the story of a blue heeler (Australian cattle dog) mix named Shep who lived with his humans on a ranch in Wyoming. During his daily perambulations, Shep often chased and killed small animals. But one fateful September day in 1981, Shep caught and killed a black-footed ferret. Biologists were shocked because the black-footed ferret was largely regarded as extinct. Shep had done an enormous service by alerting scientists to the existence of black-footed ferrets; Shep had, at the same time, killed one of what turned out to be a global population of 16 individuals. Shep's story encapsulates the promise and the peril of dogs in relation to wildlife conservation.

The impacts of dogs on wildlife are diverse, complex, and in many ways, still very poorly understood. Because of our commensal relationship, humans and dogs live mainly in the same places; the distribution of dogs mirrors the distribution of humans. This means that dogs are almost everywhere on the planet. And there are a lot of dogs! There are more domestic dogs than all other canids combined. Dogs are referred to by biologists as a "subsidized predator": they receive resource inputs beyond what would be provided by the ecosystem. In other words, because dogs are given targeted human subsidies (we feed them on purpose), they can live in ecosystems at densities beyond what the ecosystem, by itself, could support, and can thus have an outsize impact on other species within that ecosystem.

Consider some of the myriad ways in which dogs can harm wildlife and natural ecosystems. Dogs chase and catch various animals, including small mammals such as rabbits and prairie dogs, birds (especially ground-nesting birds like wild turkeys), lizards, and snakes. Unlike their wild cousins, dogs are inefficient hunters—they often don't catch what they chase. But chase they do. Because pet dogs are well-fed, they have energy reserves to chase as much as they want. Having to flee from dogs uses up precious energy that critters need for other activities like finding food.

The effects of dogs on wildlife are often less visibly obvious than outright chasing of prey, but even more insidious. Dogs are agents of disturbance and create landscapes of fear. Fear alters the physiology, activity patterns, and habitat use of wildlife. Studies have found that wildlife activity is significantly altered by the presence of dogs. Trails used by dogs become corridors of fear within natural areas. One study, for example, found that deer distanced themselves from trails by at least 100 meters. If you consider that a trail might cut through miles of open space, you now have a corridor of space several miles long and 200 meters wide that is unusable by deer. And this, of course, is assuming that all the dogs are on leash.

We must, in fairness to dogs, also mention the expanding and exciting beneficial role of dogs in conservation efforts. Because dogs can maneuver through dense undergrowth more quickly and efficiently than human biologists and have superior olfactory skills, dogs can help biologists locate and conduct population surveys of endangered animals. Dogs can be trained to detect the scat of a particular species and have been used to locate scat of giant armadillos and giant anteaters in Brazil, grizzly bears in Canada, a rare species of kit fox in the U.S., and kiwi birds in Australia. Dogs also assist humans in projects to control or eradicate invasive species. In Wales, a cocker spaniel named Jinx has been trained to help protect seabirds. The coastline of Wales is critical breeding ground for many species of seabird, including the Manx shearwater, which nest underground in burrows on the islands along the coast. Jinx is called a "biosecurity dog": he has been trained

to sniff out brown rats, who are proliferating on the islands and who feast on seabird eggs, chicks, and even adult birds.

See also Hybrids; Olfaction; Working

Co-pilot

Like so many things dog-related, bumper stickers tell us a lot more about people than they do about dogs. Dog-themed stickers can be expressions of humor, often with a jab: "My dog is smarter than your honor student," "Wife and Dog Missing. Reward for Dog," and the ever-popular "Honk If You Can See My Wiener." Stickers express advocacy: "Adopt don't shop." "Who rescued who?" (a formulation irksome to grammarians). They repurpose religious sentiment or offer an irreverent play on words: "Dog is my co-pilot." And sometimes we bring our dogs into human politics: "My dog ate Mike Pence. You're welcome."

Coprophagy

Coprophagy is the scientific term for eating poop. As a veterinary diagnosis, a dog who spends too much time eating his own poop or the poop of other dogs may be suffering from "coprophagia." Excessive poop eating can be a sign of physical or mental illness, and guardians of highly poop-obsessed dogs should seek professional help. Admittedly, it can be hard to draw a line between normal canine behaviors and behavioral pathologies. Even normal snacking on poop is considered repellent by some dog guardians, and tolerance for the behavior can be low. Yet some poop-eating, of course, is part of the normal behavioral repertoire of dogs. The consumption by dogs of human feces is part of our shared evolutionary background and is natural, if somewhat disgusting when performed by a pet dog with whom we share ice cream cones.

Among free-roaming dogs, eating human feces is often an important source of nutrition. A study of dogs living in rural Zimbabwe found that human poop was their fourth most common food. Human feces are consistently available and provide a relatively high-quality source of protein—better, for example, than the *sazda* (maize porridge) that many free-roaming dogs are fed.

See also Diet; Food resources

Counter surfing

Circa late 1990s, early 2000s. Precise origin of the term is uncertain. "Counter surfing" describes a particular

form of canine food acquisition behavior that occurs within the human home and involves quickly snatching edible items from kitchen surfaces using snout, paws, or whatever body part is available. In some cases, a counter-surfing dog will perform incredible feats of athleticism, launching the entire body onto high counters for food retrieval. The behavior is usually surreptitious and involves some planning on the part of the dog—a dog waits until all humans have left the area or until an unsuspecting human has turned their back for a moment. Counter surfing is one among many words and phrases used by humans to moralize the behavior of pet dogs, and to circumscribe the nature of our shared ecosystem: all food within the human home belongs to the human, unless directly earmarked, by the human, as intended for the dog. Dogs must only eat food labeled by humans as "dog food." A dog trying to access food resources within the house is "stealing."

See also Diet; Food resources; Kibble

Crating

Crating is a verbification of the noun "crate," a box made of wood slats or latticed plastic and used for transporting or storing goods. Within contemporary dog culture, a crate is a box used to transport and store dogs, and crating is the practice, increasingly common in some twenty-first-century dog-keeping cultures, of confining a pet dog to a small cage within the home. Crating has several ostensible functions: it is used for house-training puppies; to physically constrain an adult dog when not being supervised by a human or when interaction with a human is undesirable; and to prevent dogs from destroying items that humans value and label as personal property. Crating is disparagingly called a convenience practice by critics—which is to say, it may make life easier for dog owners, while at the same time making life more difficult for dogs. Yet crates are undoubtedly useful in some circumstances. They are used, for example, to safely transport pet dogs out of areas hit by flood, hurricane, fire, or other natural disaster.

See also Fake turf; Pet

Cultural attitudes

Human attitudes toward and treatment of dogs vary enormously across cultures and historical periods. Even within cultures, attitudes toward and treatment of dogs vary widely from one place to another, one home to another. Indeed, the only safe generalization is that you can't generalize. (This is a trap that even your trusty guide here has fallen into occasionally. When reference has been made to "pet-keeping countries," for example, this is just a broad brushstroke. Many individuals within contemporary American culture neither keep pets nor see the point of keeping pets—and may even consider the practice morally or hygienically offensive.)

One generalization can be safely proffered: where there are people, there are dogs. And where there are dogs, there are (diverse) human practices and beliefs about what kind of being dogs are, what kinds of relationships with dogs are morally or spiritually appropriate, and so on. A second general pattern is that in many places, dogs are viewed as companions by at least some people, though what this means in practice varies. A third general pattern is that in a great many places, dogs partner with humans in doing certain kinds of work, often related to agricultural practices and food acquisition.

Here are some of the attitudes and practices that might vary across time or place: 1) feeding regimes: Are dogs provisioned or not provisioned? Are they fed high-quality food items, or scraps and garbage and low-quality foods? 2) spaces where dogs are allowed and not allowed: Do dogs come indoors? Where do dogs spend most of their time during the day? Do dogs share our beds? 3) familial relations: Are dogs considered part of the nuclear family? Are they treated like children? 4) human consumption: Are dogs on the human menu? 5) burial: Are dogs given ceremonial burials? If so, are they buried alongside people? Or in separate areas? 6) work: Are dogs laborers or pets or both? 7) treatment: Is cruel treatment acceptable? What counts as cruel treatment? (Beating? Vivisecting? Forced breeding? Making dogs race or fight each other for human entertainment? Keeping dogs inside? Keeping dogs outside?)

See also Abundance and distribution; Companion animal; Epitaphs; Hayırsız Ada Dog Massacre; Pet; Stray; Streeties

Cur

"Cur" sounds an awful lot like "grrrrrrr." The word first appeared in the thirteenth century and is probably derived from Old Norse *kurra* or Middle Low German *korren*, both meaning "to growl." (Can you find other dog onomatopoeias in the English language—and other languages, too? "Bark," "howl, the "woofer" in your stereo, perhaps "zoomies"?) Like many dog-related words, "cur" carries echoes of social discrimination: "cur" was used to refer to a "low-bred man" and to a churlish, curmudgeonly person—one who is snappish and snarling.

See also Dog; Hound; Mutt

Cursorial

Dogs, like other canids, are *cursorial* animals, from the Latin *currere*, "to run"; they have long legs that make them efficient runners (unless they have been artificially bred by humans to have super short legs). Cursorial animals are adapted either to running fast, like cheetahs, or to running at a steady pace over long distances, like wolves and dogs and humans. Dogs walk, trot, gallop, and sometimes canter. Trotting is dogs' favored form of locomotion.

As in humans, there is great variation in dogs' athleticism. Like their human athlete counterparts, some dogs are built like runners, with lithe, long bodies, while some are most assuredly *not* built like runners. Good runners are not only fit, but also have an economy of movement and gait that makes them able to cover ground efficiently. A study on the effects of domestication on locomotor gait and economy found that more "wolf-like" dogs, including the northern breeds, such as Alaskan malamutes and Norwegian elkhounds, have greater aerobic economy—they expend less energy per stride—than breeds whose physical bodies are less like their wild canid relatives'.

One of the key welfare problems for dogs who are kept as pets is the lack of opportunity to run. Instead, many dogs are "walked," which typically means being attached to a collar and leash, going at a slow pace, and traveling in a straight line. Some dog guardians go to great lengths to help their dog access places where leashes are optional and dogs can let loose and run, lope, stop and sniff, dart, and zoom to their heart's content. Being unleashed can bring great joy to dogs. A 2013 study found that dogs likely experience what is known as a "runner's high," which is thought to be caused by the release of neurotransmitters, including endocannabinoids, into the bloodstream during hard running effort. The "high" is an evolutionary reward for staying fit.

See also Leashes

Cynanthropy

Have you ever wished that you were a dog? Maybe wished it so hard that you felt your nose elongating and dark hair sprouting all over your body? If so, it seems you aren't alone. In fact, some people wish so hard to be a dog that they become one, at least in their imagination. *Cynanthropy* (from the ancient Greek *kúōn*, "dog" + *ánthrōpos*, "human") refers to the ability to shape-shift into the form of a dog or weredog and has been an important piece of folklore in many cultural traditions. Cynanthropy also appears as a superpower in some contemporary gaming circles, granting individuals the supernatural ability to assume the form of a dog, often through the power of a full moon. Clinical cynanthropy (sometimes spelled "kynanthropy") is defined, in psychiatric circles, as a delusional state in which a person believes himself or herself to be a dog. A 2022 report in the medical literature described the case of a 28-year-old single male diagnosed with cynanthropy two years after being bitten by a dog. Apparently, after COVID lockdown he began excessively researching dog bites on the internet and began increasingly to fear that he was being transformed into a dog. He began walking on all fours and barking, and obsessively checked his appearance in a mirror. A course of antidepressants prompted his recovery.

Cynoctone

In 1902, the French animal protection organization Assistance aux animaux generously donated a cynoctone to the police's dog detention center. Put simply, the cynoctone was a device used to kill large numbers of dogs at once. Caged dogs were lowered into an underground chamber, which was then filled with mixture of carbonic acid and chloroform. According to an early description of the invention, the gases in the chamber caused death by asphyxiation within six minutes; when brought back up to ground level, the cage would contain only corpses. The cynoctone was considered a wonder of modern technology: a machine that could apply principles of modern scientific progress to the slaughter of unwanted canines in cities. It was part of a larger movement toward what has euphemistically been called "humane euthanasia." Unfortunately, we now know that asphyxiation is a horrible way to die.

At the turn of the nineteenth century, bustling metropolitan areas in the U.S., Britain, and Europe—cities like Paris, London, and New York—had large populations of free-roaming dogs. As cities changed and human attitudes toward dogs changed, loose dogs began to be categorized in ways that signaled their exclusion from the community. Consensus emerged that stray dogs should be rounded up and killed; they were a nuisance, a danger, an unsettlement. They carried rabies, they were dirty, they made a lot of noise, they bit people and their pedigreed dogs. Increasingly sharp lines were drawn between pet and stray, and between purebred and mongrel; the lines were traced over and over until no one could confuse these categories of dog.

At first, efforts to rid the streets of stray dogs involved hunting them down and killing them on the spot, using sticks or rocks or ropes or whatever was available. Over time, this brutal and public killing began to rub against our tenderer impulses. Instead, dogs were rounded up and taken to centralized locations ("the pound") where they could be disposed of more efficiently and less publicly. The killing gradually became less transparent to the public, and the narrative surrounding the killing of dogs shifted from "get rid of the pests" to "stray dogs are better off dead." Killing became a work of compassion. Because the dogs *had* to be killed, the humane movement focused its attention on killing the poor waifs as painlessly as possible. The cynoctone was one iteration in this history of humane euthanasia. Before the cynoctone, there was lynching, drowning, clubbing; after the cynoctone, we turned to gas chambers and then, finally, to what is fondly called "blue juice"—the injection of sodium pentobarbital into the vein as the ultimate painless way to kill dogs who don't fit into the right human social categories.

See also Shelters; Stray

Defecation

Poop typically refers to the actual pile of brown stuff. Defecation is the physiological process of creating a pile of brown stuff, but it is also far more than this. Defecation is an important element in canine social behavior. Poop contains olfactory information for other dogs. Like other canids, dogs sometimes mark territory by defecating in a certain place. Unlike humans, who generally prefer to poop in out-of-the-way, private places, dogs like to poop out in the open, where the message will be prominently available to others.

See also Anal glands; Ground scratching; Poop laws; Urine

Diet

Although dogs belong to the taxonomic order Carnivora (placental mammals who have specialized in eating flesh), and have the dentition to tear and chew flesh, they are dietary generalists and can and do eat a wide variety of foods. Indeed, being highly flexible in whom or what they will eat is one of the things that has made dogs so successful. Dogs' diet varies considerably from one ecosystem to the next and depends on the availability and type of human-derived materials (also referred to as anthropogenic food resources), the types of prey available and how easy or hard they are to catch, and competition from other species for the same food sources. Free-roaming dogs tend to eat a lot of fruits and other vegetation, human-derived materials, and small mammals. Also, to a lesser extent, they forage on fish, birds, reptiles, amphibians, insects, and what might be classed as "non-food" items (cardboard packaging). In some areas, dogs may eat a narrow range of foods, because that's what's available to them. For example, in India, dogs mainly eat human-derived materials (garbage, feces, scraps) and vegetation; in Zimbabwe, they mostly eat mammal prey; and in the U.S., they primarily eat food provisioned by their human guardians.

The diet of homed dogs is highly variable and depends on what the guardian of the dog has decided is most nutritious, cheapest, easiest, or has the cutest packaging. Many homed dogs are fed kibble or canned foods, designed and processed by dog food manufacturers. These foods generally contain a protein source—often the rendered parts of slaughtered animals that humans find disagreeable and don't want to eat ourselves—combined with grains, vegetables, or fruits. Whereas free-roaming dogs may sometimes suffer from insufficient food, homed dogs often suffer from excess. Obesity is an enormous public health epidemic among pet dogs. Mirroring patterns seen in humans, over 50 percent of pet dogs are overweight or obese.

The question of what diet is best for dogs is highly contentious among humans who keep dogs as pets. Some have argued that dogs should eat like wolves. But the comparison to wolves is problematic in the realm of diet. For one thing, dogs don't get nearly as much exercise as their wild relatives. Perhaps more importantly, dogs have evolved to eat differently, because of their close association with humans and the sharing of food between humans and dogs. Domestication has altered dogs' digestive system. Their microbiome is different from wolves', and they have adaptations, including the alpha-amylase 2B gene for digesting starches, that allow them to process a wider range of foods.

See also Food resources; Kibble; Taxonomic classification

Digging

Why must dogs dig holes in our perfectly landscaped gardens? Well, dogs might have their own sense of what's beautiful. And, perhaps more importantly, digging just seems to bring them joy. Digging is an instinctive behavior, and dogs may feel the need or desire to dig even if it doesn't serve any immediate purpose. The behavior may be related to hunting prey, especially going after burrowing animals and insects. It may also relate to denning. Studies of free-roaming dogs have included observations of denning. Dogs dig holes to bury stuff, like bones. This is a smart way of caching food for later. Dogs who dig may be looking for items stashed by other digging dogs. Sometimes digging may be an attempt to escape, by digging under a fence or other barrier. Sometimes dogs will dig down into sand or dirt to create a cool space to rest on a hot day. Some dogs may be bored, and digging may provide an outlet for creativity or a way to burn off nervous energy. Some dogs may just be trying to dig their way to China.

Digging behavior has been developed in some breeds of dog through selective breeding. For example, terriers, who were bred to hunt critters who live in holes (like rats), may be especially driven to dig. Whatever the reason, many dogs like to dig, and find digging satisfying. Unfortunately, digging—like many other natural dog behaviors—is something for which dogs are often scolded or punished in human environments. To provide acceptable outlets for digging, some dog parks have designated digging holes, and some dog guardians will provide special dig-friendly places like a sandbox in their yard.

See also Joy; Selective breeding

Displacement behavior

The term "displacement behavior" was coined by ethologists as a way of describing a normal behavior that seemed out of context, or "displaced." Animals engage in displacement behaviors when in a state of inner conflict or anxiety—perhaps they are curious, but also afraid. The individual tries to achieve a sense of control and calm by performing an activity that feels safe. Some displacement behaviors seen in dogs are yawning, lip licking, tongue flicking, grooming, sniffing at the ground before meeting another dog, urinating, and picking up an object like a toy or stick and carrying it.

Human displacement behaviors include fidgeting with your hair when you are talking to someone you find sexy, or scratching your head when you can't decide what to do.

See also Yawning

Dog

"Dog" is a biological designation (quadruped of the genus *Canis*), but it is so much more. Oddly, for a word that is used millions of times a day in English-speaking countries, its etymology remains one of the great mysteries of the English language. Best guesses suggest that "dog" comes from the Proto-Indo-European root *kwon* (also the root of canid and canine). By at least the twelfth century, "dog" was used in reference to persons, and not in a complimentary way. The various uses and meanings of "dog" point toward a long history of negative attitudes toward dogs. Used figuratively, a "dog" is a worthless or contemptible person, an unattractive woman (sprinkling some sexism onto our speciesism), an undesirable or inferior piece of merchandise. To "dog" or "hound" someone is to be a source of constant, unrelenting irritation; "going to the dogs" suggests impending ruin.

See also Cultural attitudes; Cur; Mutt; Hound

Dogor

It is a dog? Or . . . ? Dogor, which means "friend" in the Yakut language, was the name given to the 18,000-year-old remains of an early canine specimen found in 2018

in Siberia and remarkably well-preserved by permafrost. Dogor appears to have been about two months old when he died—he still has his milk teeth. Sequencing of DNA extracted from one of Dogor's rib bones puzzled scientists, who couldn't decide whether he was a dog or a wolf, or perhaps a transitional species. Eventually, scientists concluded that Dogor was a wolf who lived during a period when early dogs were beginning to evolve. Even with the Dogor mystery solved, questions about dog evolution and domestication remain in flux and are keeping scientists very busy. If Dogor had been a dog, he would be the oldest ever found to this point. The so-called Bonn-Oberkassel puppy, whose remains were found in a quarry in Germany, has been aged at 14,200 years old, and is currently the oldest identified dog specimen. But there are probably proto-dog remains that are at least as old as Dogor, and our story of dogs' history will surely continue to change as new discoveries are made.

See also Domestication; Paleofeces; Teeth

Domestication

Domestication is a biological process that has led to the formation of unique human-animal relationships. Domestication is not the same as taming. Taming refers to the habituation of an individual animal to the presence of humans. Taming does not alter DNA. Domestication, on the other hand, involves genetic changes that are heritable, and is something that happens to a whole species. Taming was likely part of the domestication process for dogs: tame wolves became, at some nebulous point or several points, domesticated dogs. Dogs are one of only 14 large mammal species to have undergone domestication.

The domestication of dogs was arguably one of the most important events in both human and canine history. And yet that's about all we can say without dipping our feet into one controversy or another. It happened such a long time ago, the only historical record is fossils, and it was an incredibly complex series of events. There is much we still don't understand, and new archaeological data and genetic analyses are constantly updating and altering the story. We can review in broad strokes how dog domestication may have gone down, but take everything here with a grain of salt—because new evidence will certainly have emerged, and the scientific consensus evolved, by the time this book is in print.

The domestication process for dogs likely took place over thousands of years (it is still underway), beginning roughly 40,000 years ago. Some scientists push the date back as far as 100,000 years; others believe the evidence favors a shorter timeframe of 15,000 years ago. Domestication probably took place on the Eurasian continent, but perhaps in more than one location on the continent. Dogs are descended from at least one, maybe several, precursors to today's gray wolves and have undergone a process of natural and artificial selection that has affected their morphology, cognition, behavior, and metabolism. One possible scenario is that some wolves were more predisposed than others to tolerate the presence of humans and felt lower levels of fear. The wolves who chose to live in proximity to humans had fitness advantages such as better access to food and a level of protection from predators.

At some point, humans started to selectively (though unconsciously—they wouldn't have known modern genetics) breed dogs for certain functions, such as pulling sleds. At some much later point (at least 2,000 years ago, perhaps more), humans began selecting and breeding for highly specialized physical and behavioral traits, in a process that we now call artificial selection. Though many current dogs are the product of deliberate breeding for certain traits, many are not. Dogs still mate and reproduce outside human channels and live semi-independently of humans. Some dogs, such as the dingo and the New Guinea singing dog, may have undergone early stages of domestication, but have been free of direct human selective pressures for some time.

You often hear generalized statements such as "humans domesticated dogs." But bear in mind that domestication is a profoundly complicated and synergistic process involving human-directed changes as well as adaptations to specific and perhaps novel ecological niches. Humans cannot—sorry to say—take full credit for domesticating dogs. Dogs did a lot of the work themselves, and nature had a large hand in it, too. And just as dogs have evolved through their close interrelationships with us, so too have humans been indelibly shaped by our long history with dogs.

See also Artificial selection; Attachment; Breed; Landrace; Neoteny; Paedomorphism; Puppy-dog eyes; Selective breeding; Wolves

Dominance

Dominance is one of the most misunderstood and most consequential concepts in the realm of contemporary human-dog relationships. Dominance is often confused with or conflated with other things, including aggression, leadership, obedience, and punishment.

Simply put, dominance is control over the behavior of a conspecific (a member of the same species). Dominance, perhaps counterintuitively, is an evolutionary strategy for reducing conflict, which it does by establishing rank, and by creating clear signals of rank. Strategies for achieving and maintaining dominance sometimes involve physical interaction; dominance can also be communicated using body language, facial expression, and olfactory signaling, among other things.

Dominance and submission are linked concepts, used to understand and describe the behavior of social animals for whom intragroup, or within-group, conflict is costly. Social animals must work together to survive, and survival depends on cooperating, negotiating, and retaining peaceful relations. Fighting takes time and energy and can lead to injury or death. Social animals who are aggressive (e.g., wolf, dog, human) have evolved various strategies for reducing conflict, and dominant/submissive relations are one of these. Dominance and submission are extremely important in wolf packs; dogs also understand and use dominance and submission, but the behaviors don't function in precisely the same way for dogs as they do for wolves.

Dominance-based training—colloquially referred to as "I am the boss of you!"—has held a strange appeal within dog-training circles over several decades. (Strange because it is both scientifically and ethically flawed.) Although dominance-based training is a loose and imprecise designation, it generally reflects the idea that to train well we need to be in a position of power, and we gain and maintain this power through brute force and intimidation. Training methods include the so-called alpha role, which involves pushing a dog onto her back and pinning her there, usually by holding the throat until she stops struggling against you; "scruffing" (forcefully grabbing hold of the loose skin of a dog's neck); and grabbing and holding a dog's snout. Contemporary evidence-based training has moved beyond these fraught methods.

See also Aggression; Appeasement; Submissive behavior

Dreaming

Do dogs dream of chasing rabbits? Almost certainly, or at least they dream of chasing something or someone. Scientific understanding of animal dreaming is accumulating and supports the idea that dreaming is a neural activity spanning a diverse range of animal species. When dogs dream, they are likely replaying, rehearsing, or running through behavioral simulations of canine reality. A question we cannot answer yet is whether a dreaming dog ever successfully catches her leporine prey.

Ears

The dog's ear is a wonder of anatomy. The pinna, or ear flap, has 18 muscles which allow the ear to raise up and lower down, rotate and twist, perk and droop. By comparison, the human ear has only 10 muscles, 4 of which are vestigial and serve no real purpose. Dogs have exceptional hearing—far better than our own. A dog's ear is like a funnel for sound; dogs can move their ears separately, so one can point forward while one is rotated backward or sideways, the better to take in auditory information from multiple directions. Head tilting might be a dog trying to hear better by changing the angle at which sound waves are hitting the ear canal.

In addition to collecting sensory information, ears are used to communicate mood and intention. Ears pulled back communicate appeasement, ears standing up and pointing forward communicate alertness and interest, ears flattened down communicate an agonistic response, while ears held sideways express a state of inner conflict.

Dog ears come in variety of shapes and sizes: pricked (erect and V-shaped), floppy or dropped, semi-erect, and lobate, to mention just a few. It has been hypothesized that some of these ear shapes might compromise a dog's ability to communicate with other dogs and could also potentially affect the acuity of hearing. Extremely long and droopy ears, such as we see on the basset hound, cannot rotate very easily, and cannot prick or flatten with nearly as much nuance as a "regular" (wolf-like) dog ear. Surgical altering by humans of dogs' ears—a practice known as "cropping"—also effects communication by forcing the ears into an unnaturally pointy shape with reduced range of motion. A dog with permanently pricked ears, such as those seen on the Doberman pinscher, is rather like a human who has had too much Botox and whose face looks frozen in place.

One additional function of ears is temperature regulation. When a dog gets hot, blood vessels in the ears can expand, enabling more blood to flow closer to the surface of the skin. Hot ears on a dog can be a sign of fever.

When puppies are born, their ears and eyes are closed, an evolutionary strategy that protects these delicate and essential sensory structures from injury. Both will be open by the time a pup is about 14 days old.

See also Communication; Facial expressions

Ecological niches
Ecological niche refers to the various factors that play a role in a species' survival, including food, shelter, competition, and climate. Humans and dogs don't have the same ecological niche, but we have overlapping niches. We can exist in the same place together but fill slightly different and mostly complementary roles. It is often said that the ecological niche of the domestic dog is the human home. But this is an oversimplification, and a human-centered way of describing things, assuming that the niche is created by us and we give dogs permission to live in it. It also overlooks the fact that the majority of the world's dogs don't live within human homes.

See also Abundance and distribution; Activity patterns; Conservation impact of dogs; Diet; Food resources; Pet; Stray; Streeties

Electricity
Who would have guessed that dogs were involved, albeit nonconsensually, with the development of electricity? Let us go back in time to a famous feud between Thomas Edison and George Westinghouse in the late nineteenth century, over two competing systems of electric lighting. Edison's direct current had trouble traveling over extended distances; Westinghouse's alternating

current didn't have this problem, and it drove Edison crazy with jealousy. Edison began an aggressive campaign to discredit Westinghouse's alternating current, demonstrating that it was dangerous by holding public executions by electrocution not only of a well-known circus elephant named Topsy, but also of various stray dogs whom he had purchased for 25 cents each from neighborhood boys who enjoyed the sport of rounding up loose dogs. Edison also spent time in his laboratory in New Jersey exploring what would happen if one attached electrodes to calves and horses—how long would it take the animals to die under varying strengths of current? Try not to think about that the next time you turn on your lights.

Emotions

It is hard to imagine, if you have shared close friendship with a dog, that the question "Do dogs have emotions?" was ever entertained by serious scientists. But it was, and it took until just the last few decades for animal emotions, including dog emotions, to become a major focal point of scientific research. Now that the winds have shifted, people are busy trying to understand the complex emotional experiences of dogs, and, more importantly, how we can harness this knowledge to help dogs living with us to be happier and more emotionally healthy.

Emotions are psychological phenomena that help animals manage and control behavioral responses to their environment. Fear, for example, elicits an adaptive behavioral response (run!). What emotions do dogs experience? For starters, dogs likely experience fear, anger, sadness, happiness, joy, shame, embarrassment, resentment, rage, love, jealousy, pleasure, disgust, grief, and despair. Some questions that we cannot now answer with certainty (and probably never will): How are their inner experiences of these emotions different from our own? Are there emotions that are unique to dogs (for which we have no label, since we are incapable of even conceiving what these might be)?

Although inner states are, by definition, private and subjective, we can often infer the inner states of other animals by using what are called behavioral correlates. When an animal is doing X, they are likely feeling Y. The emotions of dogs are often evident in their facial expressions and body postures and can also be inferred from how they make decisions. One methodology used for inferring emotional states relies on testing what is called "cognitive bias." Mood states are thought to bias decision making, with negative moods states leading an animal to be pessimistic about outcomes, to view the glass as half empty rather than half full, and to make decisions based on that pessimism. One study of cognitive bias in pet dogs found that dogs who exhibited high levels of separation anxiety were less optimistic about the possibility of finding food in a bowl placed across the room than their happier peers.

Dogs and humans—and many other animals— experience a phenomenon called emotional contagion. Emotions are infectious: emotional states spread from one person to the next, from one dog to the next, and even from person to dog and vice versa. Studies of emotional contagion in dogs—all of them conducted with pet dogs—have come up with a range of findings:

dogs are highly sensitive to the emotional states of their human guardian, and the longer they live with a person, the more they "catch" that person's feelings. In particular, if a dog's guardian is anxious, the dog is likely also to feel anxious. Female dogs show more contagion than male dogs.

See also Communication; Eureka!; Grief; Guilt; Inequity aversion; Jealousy; Joy; Love; Zoomies

Epitaphs

Humans have ceremonially buried, mourned, and honored the lives of dogs for thousands of years. Some of the most visible markers of our commemoration are the epitaphs written for dogs and often placed on their grave. More epitaphs for dogs are found in the historical record than for any other animal. These epitaphs give us a window into how deep the human affection for dogs can run and reflect practices of boundary setting between dogkind and humankind.

An Egyptian dog named Abutiu (also spelled Abuwtiyuw) is one of the earliest known domesticated dogs to be given an epitaph—indeed, one of the earliest documented animals to have his or her name recorded. Dated to about 3100 BCE, an inscribed block of white limestone in Abutiu's tomb tells us that the beloved dog of a king has died. The king wanted to be sure that the soul (or *ka*) of his dog would reach the afterlife and would be waiting for him when his own death came. The king ordered that the dog be buried in a coffin from the royal treasury, lined with red linen. Abutiu was likely a *tesem*, a hunting dog who might have looked like a modern-day greyhound.

The Greeks and Romans memorialized certain dogs and gave them burials and tombstones. On a tombstone with marble relief, erected sometime around 100–200 CE and found at Salernum, a lyrical poem honors the life of a dog named Patrice:

> *Bedewed with tears I have carried you, our little dog, as in happier circumstances, I did fifteen years ago. So now, Patrice, you will no longer give me a thousand kisses nor will you be able to lie affectionately 'round my neck. You were a good dog and, in sorrow, I have placed you in a marble tomb and I have united you forever to myself when I die. . . . You, sweet Patrice, . . . were accustomed to lick with your greedy tongue the cup which my hands often held for you and regularly to welcome your tired master with wagging tail.*

Lord Byron wrote an entire poem for his Newfoundland dog, Boatswain, who died of rabies in 1808. Sometimes called "Epitaph to a Dog," the poem is inscribed on Boatswain's tomb, which happens to be much larger than the tomb of Byron himself, perhaps a fitting reflection of Boatswain's superior moral character. Both tombs are at Newstead Abbey, Byron's estate. The poem begins with a description of Boatswain's impeccable moral virtue.

> *Near this Spot*
> *are deposited the Remains of one*
> *who possessed Beauty without Vanity,*
> *Strength without Insolence,*
> *Courage without Ferosity,*
> *and all the virtues of Man without his Vices.*

Byron goes on to expound further upon the nobility of his dog's character, especially in comparison to hypocritical, lustful, deceitful, and vile humans, an interesting choice of focus, considering Byron's own moral lassitude. Yet Byron touches on a common theme in human-dog relations: we often project onto dogs those qualities of character to which we aspire and at which we are most likely to fail.

Modern epitaphs can be found at any of the hundreds of dedicated pet cemeteries in the U.S. and elsewhere around the world. Academic researchers whose work focuses on dog epitaphs—and yes, there are people who specialize in this area—believe that epitaphs are a window into cultural attitudes toward pets. For instance, scholars have noted a distinct shift in the early to mid-nineteenth century in the U.S. Dogs started to be spoken of as family, and gravestones began to denote human guardians as "Mommy" or "Daddy" and dogs as "our beloved baby." Dogs were increasingly referred to using surnames ("Brownie Smith"). Another interesting shift reflected in epitaphs on dog tombstones occurred in the 1940s and '50s: the gates of heaven slowly creaked open wide enough to allow dogs. More and more gravestones of dogs had religious symbols or allusions to heaven.

See also Cultural attitudes; Domestication; Grief; Love; Pet

Eureka!

Animals are wired to find hard work rewarding; reward, in the form of neurochemicals that create positive emotions, is nature's way of reinforcing behaviors that help organisms survive. Researchers have explored various angles of the "why work is rewarding" idea. A large body of research on what is called "contra-freeloading," for example, shows that given a choice, animals will work for food or some other reward rather than take the reward for free; they don't want to be freeloaders. Hard effort, both physical or cognitive, is intrinsically satisfying and leads to positive psychological states. Like us, other animals experience positive emotions—what we humans often refer to as the "feeling of accomplishment"—after putting effort into solving a problem. Dogs also feel satisfaction when they discover something new. In an experimental setup in a lab, dogs who successfully learned to solve a novel problem experienced a burst of excitement. Researchers called this the "eureka effect."

See also Emotions

F **acial expressions**
It is often said of humans that the face is a window to the soul. We unconsciously make intuitive judgments about other people's trustworthiness simply by looking at their face—and these judgments are fairly accurate. Facial expressions are equally important to dogs. A dog's facial expression is a mosaic of communication points: ears, eyes, mouth, and so forth. Together, these points offer a clear picture for other dogs (and for humans who are paying attention) of mood and intention.

Here are a few mosaic pieces. **Eyes:** In dog-to-dog communications, staring is often a threat, while averting eyes is a calming signal. (In dog-to-human communication, staring can play a different role: dogs will gaze for long moments into the eyes of humans with whom they have a strong bond and are communicating affection.) "Soft" eyes indicate a relaxed, non-threatening inner state; "hard" eyes indicate tension. In agonistic situations, dogs may open their eyes very wide, exposing the sclera, or whites—often called "whale eye." Pupil dilation and blink rate add even more nuance. **Mouth:** The labial commissures (corners of the mouth) are drawn forward into "short lips" during agonistic displays and are pulled back ("long lips") to communicate stress. **Ears:** forward (interest), back (appeasement), flattened (fear or agonistic response), really flattened (really scared), sideways (inner conflict).

Although dogs read each other's inner states by looking at faces—alongside the larger composite picture of body posture, vocal cues, and so forth—it seems that dogs are less interested in human faces than in dog faces. One small study showed that dogs' brains light up when a dog's face comes into view, but not when a human face comes into view. Nevertheless, dogs can and do read emotions from human faces, and in one experiment showed increased heart rate when looking at human faces expressing anger, happiness, and fear—all states of emotional arousal.

Research has found that humans are pretty good at identifying dogs' moods and emotions based on their facial expressions. With training and attention to nuance, humans can become quite adept at using dog facial expressions to understand their emotional states. This skill can be particularly valuable in helping us understand when a dog is feeling discomfort or pain, and in knowing when a dog is open to being approached

and when we should be cautious. Some scientists have speculated that selection processes during evolution may have led to functional equivalencies in facial communications—in other words, dogs and humans make some of the same facial expressions, even though we don't share the same facial anatomy. Humans lift the inner eyebrow to express sadness. Although dogs don't technically have eyebrows, they can move a muscle above the corner of their eye and give the appearance of raising an eyebrow. A research team is developing what's called the DogFACS, based on the model of Paul Ekman and Wallace Friesen's Facial Action Coding System, which is a comprehensive system for describing every conceivable facial movement and linking them with underlying emotions.

See also Communication; Ears; Emotions; Puppy-dog eyes

Fake turf

One of the products gaining traction in the pet product market over the past decade is the doggie equivalent of a kitty litter box. Dogs, of course, don't bury their poop like cats, and would have no interest in a box filled with tiny pieces of scented clay. They do, however, find grass to be an acceptable substrate for relieving themselves. And indoor fake grass litter boxes for dogs are now readily available for purchase. Some dog guardians use the fake turf during house-training of puppies; others use the little square of turf as a more permanent solution to the annoying problem of having to take a dog outside to relieve herself. There is now a population of pet dogs that never leave the house or apartment, even to eliminate—the canine equivalent of indoor-only cats. This signals a significant shift in dog-keeping practices, and an intensification of the captivity experienced by some pet dogs.

A potential problem with doggy litter boxes is that dogs lose the opportunity to engage in a whole suite of elimination-related behaviors, which are either impossible or, at the least, extremely uninteresting on a strip of plastic grass: they can't sniff to gather information from the pee and poop of other dogs and can't use their pee and poop to send olfactory messages to other dogs. In other words, they lose any semblance of a social community.

See also Defecation; Ground scratching; Olfaction; Poop laws; Sniffing

Feral

A feral animal lives in the wild but is descended from a domesticated ancestor. Feral dogs are the same species—*Canis lupus familiaris*—as toy poodles riding around in handbags, though the similarities may be hard to spot. Feral dogs live on their own, and although they typically rely on anthropogenic food resources, they fear humans and avoid close interaction. The process of feralization refers to changes in individual dogs, rather than to changes at the population or species level. A pet dog can become feral over time; a feral dog can also become a pet dog, though adjusting to a human home environment can be challenging.

How many feral dogs are there in the world and where do they live? It's impossible to say. Nobody has collected this data (it would likely be difficult to gather in a comprehensive way), and the line between stray and feral is blurry and shifting. Global assessment is out of reach, but here is one snapshot. India has a large population of free-roaming dogs—somewhere between 45 and 60 million dogs. Loose estimates suggest that only 5–10 percent of these dogs are truly feral.

See also Nomenclature; Pet; Stray; Streeties

Flea control

For such tiny creatures, fleas have played an outsize role in the history of dogs and dog-human relationships.

Dogs have always lived closely with fleas, and in living closely with dogs, humans have also faced down this common enemy.

Parasite control, and especially the control of fleas, has been identified by historians as a key moment in dog-human history: the development of effective methods to control fleas during the early to mid-twentieth century made it far more inviting to keep dogs in the house, and was probably one of the factors influencing modern dog-keeping practices in which dogs live fully inside our dwellings. In 1929, references began appearing to dogs' bedding as a potential source of fleas, with the recommendation that bedding be taken outside regularly, shaken vigorously, and left in the sun for a few hours. For on-animal control of fleas, dog owners were advised in these early days to dust their pets with derris powders, powdered naphthalene or pyrethrum, or pulverized mothballs, which would "stupefy" the fleas, who could then be collected as they dropped off the animal and could be burned. Dogs were also washed with carbolic acid soap. After World War II, dichlorodiphenyltrichloroethane (DDT) was considered safe enough to be used in the home on pets and humans alike and was recommended for control of fleas. Although effective against fleas, many of these early treatments had the unfortunate side effect of making dogs and sometimes also people very sick.

Nearly all veterinary practices now recommend year-round preventative treatment for fleas (and ticks—treatments tend to cover both parasites), and most products on the market have relatively low toxicity to dogs, especially compared to such chemical monstrosities as

DDT. Still, not all dog guardians comply with the industry recommendations, often out of concern for their dog's long-term health. Although modern flea control products are generally safer for dogs and people than 100 years ago, some risk-benefit analysis is involved in every human decision to preventively treat parasites in dogs, and some feel that the recommendations are unduly influenced by profits. According to one estimate, 4 billion is spent on prescription flea treatments, and $2.8 million on flea-related veterinary bills, each year in the U.S.

Flea infestation is most certainly unpleasant to dogs. The fleas feed on blood, and the tiny bites hurt—that's what causes dogs to scratch. Dogs infested with fleas can scratch themselves raw, particularly if they have an allergic reaction to the flea bites. Sometimes a flea infestation can be so severe that it interferes with overall physical functioning. Can fleas transmit diseases to dogs? Yes, indeed. Fleas are excellent vectors of diseases, including bartonellosis, a bacterial infection that causes fever, irregular heartbeat, joint pain, and neurological changes, among other things. Fleas can also transmit plague and typhus, but not the species of flea generally found on dogs. Fleas are tiny ecosystems of their own. Some fleas, for example, carry tapeworm eggs. A dog can get infected with tapeworm by swallowing a "tapeworm-enhanced" flea.

Let's take one moment to talk about how remarkable fleas are: they are nearly indestructible and can jump 50 times their body length to land on a passing host animal whose body heat and exhaled carbon dioxide they have sensed. These tiny wingless insects first evolved

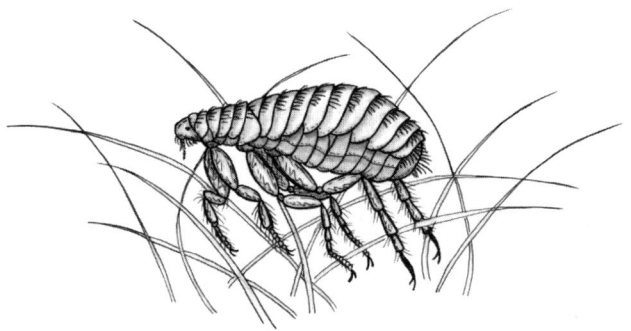

during the Middle Jurassic period and are well-adapted for feeding on the blood of mammals. The flea's body is flat, and its legs are covered in claws—both of which help it cling to and travel between the hair shafts on a furred animal's body. Earth is host to some 2,500 different flea species. The most common flea found on pet dogs in the U.S. is called the cat flea (*Ctenocephalides felis*), offering support for dogs' claims that cats are the root of all evil.

Associations between dogs and fleas are deeply grooved in the human imagination. Fleas and dogs go together like peanut butter and jelly. A scratching dog is often accused of being flea-infested, generally without supporting evidence, and being flea-infested brings with it connotations of being dirty and unkempt. Fleas appear in one of our common dog-related idioms: "If you lie down with dogs, you'll wake up with fleas." In human terms, if you hang around with unscrupulous people, you'll wind up getting into trouble. Luckily, the metaphor has little real bite: dogs are not nearly as unscrupulous as we are, so lying down with dogs is normally safe. Also, fleas don't actually live on humans—we don't have enough hair.

See also Parasites; Pet

Flews

Some parts of the dog are easy to identify: the muzzle, chest, elbow. Others you can probably guess at: loin, knee, point of rump, occiput. But where would you find the flews, the brisket, or the pastern? The flews, also sometimes called "jowls," are the thick upper lip that hangs down the sides of a dog's muzzle and covers the teeth. The flews are the part of the dog that you might see flapping in the wind when a dog sticks his head out a car window. On some dogs, such as the bloodhound, bulldog, basset hound, and Saint Bernard, the flews are especially pronounced; saliva often pools in the loose skin of the flews and can be a source of considerable slobber and/or drool. The pastern is just above the wrist and below the elbow (imagine the human forearm). The brisket is located between the forelegs and below the chest.

See also Saliva

Food resources

The feeding behaviors and diet of dogs are tightly connected with humans and human settlements. Indeed, food is perhaps what drew dogs and humans together in the first place. Nearly all dogs on the planet currently rely on anthropogenic food resources, or food derived in one way or another from humans. Dogs, as we've explored in an earlier entry, are dietary generalists and are very opportunistic when it comes to dinner. They will eat what they can get. Dogs are also very skilled at convincing humans to give them food, a behavior sometimes referred to as "begging."

Humans have been feeding dogs on purpose for millennia, perhaps for most or all of our evolutionary journey together. A cuneiform tablet from southern Iraq dating back to roughly 2000 BCE, which provides administrative details about the management of an ancient city, includes a record of the apparently quite generous allotment of food distributed to "dogs of the palace." Contemporary pet dogs are often exclusively fed specialized and expensive "dog food," even while table scraps and leftovers are tossed into the garbage. This bizarre and wasteful human behavior must be profoundly frustrating to the pet dog, who must watch morosely as the tasty bits are put into a can labeled "off limits," while he is offered only a bowl of monotonous bits of extruded, processed junk the human calls "kibble."

Free-roaming dogs rely on a mix of targeted feeding and creative foraging. Although the diet of free-roaming dogs has not been studied in tremendous detail, what research we have suggests that the primary anthropogenic food source for free-roaming dogs living in urban areas is not the most obvious—garbage dumps—but rather a limited number of "reference households" or "reference individuals" who directly feed the dogs. Nevertheless, garbage is a crucial part of the diet of many dogs around the world. Dogs are fixtures at garbage dumps in many urban settlements. Garbage has been one of the most important sources of food for dogs throughout our shared history. Dogs, for their part, have played an important though not always adequately admired role as sanitation officers. Particularly before the widespread construction of public sanitation systems and organized garbage dumps, dogs would help manage waste, in turn helping to control animals who also lived off the garbage but were considered vermin, such as rats and mice. When cities started collecting garbage and moving it to their outskirts, the role of dogs shifted. Not coincidentally, the rise of public sanitation coincides with broad-scale persecution and eradication of "stray" dogs—their work was no longer essential.

Dogs also eat human feces, another service for which they are vastly underappreciated. Human feces have long been and still are a key food source. Fossilized dog poop from thousands of years in the past has been found to contain human DNA, most likely from dogs eating and digesting human poop. One contemporary study in Ethiopia found that human feces make up about 20 percent of the diet of village dogs. As unsettling as this may seem to us who live with a pet dog, human poop is a decent source of caloric energy. So, in addition to their garbage removal work, dogs should be graciously thanked for their role in managing human hygiene and public health. Among other things, the "removal" of human feces helps control infectious disease.

See also Begging; Diet; Ecological niches; Kibble; Paleofeces

Free-roaming

The loose category "free-roaming" refers to dogs who have significant opportunities to freely choose where they go and when. The term "free-ranging" is sometimes used instead, and they are interchangeable. Free-roaming dogs almost always live near humans and exploit human food resources, including garbage, waste, or directed handouts. Free-roaming is usually contrasted with "pet" or "homed," but dog researchers include homed dogs who have substantial freedom of movement within the free-roaming category, as well as the many dogs who are otherwise referred to as stray, street, village, poorly supervised, uncontained, or feral. Like "homed" and "pet," "free-roaming" is an imprecise designation. The categorization of dogs (e.g., into stray, pet, feral, etc.) is very fluid; dogs may occupy several overlapping categories and may shift from one to another over time.

Free-roaming dogs inhabit diverse ecosystems and exist across a broad spectrum of independence from humans. Some free-roaming dogs interact often and in friendly ways with humans and may even have homes with which they associate. Other free-roaming dogs are well on their way to being feral. Populations of free-roaming dogs tend to be larger in developing countries and in urban as opposed to rural areas, perhaps because anthropogenic food resources are more plentiful in densely populated human settlements.

How many of the world's dogs are free-roaming? It is hard to say exactly: as we've noted before, dogs are hard to count (they move around a lot) and can drift from one category to another (e.g., from homed to free-roaming and back again). But rough estimates put the number of free-roaming dogs at about 720 million, about 80 percent of the world's population of dogs.

See also Feral; Homed; Nomenclature; Pet; Stray; Streeties

Gazing

Gazing
Here's an interesting thing about eye contact and dogs, reflecting the effects of our long relationship: dogs have modified the functional meaning of a species-specific behavior pattern and have adapted it to dog-human communication. Dogs use eye contact differently with humans than they do with each other. In dog-dog communication, intently looking at someone is a threat. In dog-human communication, intent gazing is a request for help or a solicitation of affection. Dogs, including young puppies, seem to have a spontaneous interest in human faces and will make eye contact with humans. Dogs lock eyes with us when asking for food or attention, when they need help getting the door open so they can get outside to pee, or when they can't figure out the puzzle toy we've offered. A bonded human and dog gazing at each other can be a powerful experience—and science helps explain why. Gazing at each other has oxytocin-mediated effects; oxytocin is a hormone that enhances the formation of an affiliative bond and is what drives the emotional draw between mother and infant. In layman's terms, when we and our dogs gaze at each other, we both feel the tug of heartstrings. We feel love.

Probably the most notable thing about the canine gaze is the human obsession with it. Scientists have been particularly obsessed by "gaze-following" by dogs: if a human is looking intently at something, a dog will look, too. We've also been obsessed with dogs following human pointing gestures, in which a dog follows with his or her gaze to where a human is pointing. Somehow, even though dogs don't have fingers per se, they still know what it means when a human points at something—which is remarkable and might mean that the domestication process has given dogs tools specifically designed to communicate *with us*. Scientists have been especially interested in whether wolves follow human gaze and pointing gestures. Although dogs generally seem far more interested in gazing and pointing than wolves, a few studies have shown that wolves perform just as well as dogs at contrived gaze-following and point-following tasks, suggesting that we still have a lot to learn about both dogs and wolves and the differences between them.

Not nearly as much attention is given to when and why humans gaze at dogs—but we do. We gaze at a group of dogs playing at the dog park (spiritual uplift), at our dog as she sleeps curled in front of the fireplace (love), or at our dog as she explores the world of smell beneath our feet (curiosity and awe). We can work to gaze more often and with greater attention at our dogs. And we can practice the "soft" gaze recommended by yoga teachers, where we quiet our mind and simply open ourselves to awareness. We might notice things about our dog that we had missed before.

See also Communication; Domestication; Emotions; Love; Wolves

Greeting

Greeting behaviors are important in social species. When two individuals meet, how do they communicate their intentions? Do they come in peace? Perhaps counterintuitively, greeting behaviors are related to aggression; ritualized greeting behaviors reduce the possibility for

aggression by signaling intentions and reducing uncertainty. Canine greeting behaviors include body posture and facial expression as the dogs draw close: dogs will often present with a curved body posture that is sometimes described as banana-like, and a loose, relaxed gait. The facial expression will be relaxed but alert. When they are close enough, dogs will sniff each other's muzzle, then sniff each other's genitals, then perhaps sniff muzzles again. From there, the dogs may go on to engage in play, may go their own separate ways, or very, very occasionally, may get into a tussle.

See also Aggression; Appeasement; Communication; Facial expressions; Sniffing; Submissive behavior

Grief

Although there is an unavoidable privacy to any animal's subjective experience, there is every reason to believe that dogs can or do experience grief. Grief is almost certainly an emotional experience shared by all animals who form close social bonds of love and attachment. In humans, grief is described as feelings of fear, sadness, loneliness, anxiety, pain, panic, or yearning experienced when a loved one dies. This is likely a fairly accurate picture of how grief is experienced by dogs. Mourning is the outward, behavioral expression of grief.

Unfortunately, because scientists have until recently been reluctant to affirm and explore emotional experiences in nonhuman animals, we don't have robust scientific literature on grief and mourning in dogs. What we do have are piles and piles of anecdotal evidence. Dogs have been observed in mourning over the loss of a pup, a parent, a sibling, a canine friend, or a life-long canine housemate. Dogs also grieve the loss of special human companions. Behavioral signs of mourning include lethargy, decreased activity, decreased appetite, sleeping near places where the lost one spent time, and depression. Grief and mourning in free-roaming and feral dogs are an unwritten book—we know almost nothing.

The depth of human grief in response to the death of a dog is one of the most remarkable aspects of our interspecies friendship. Human grief over the loss of a dog can match, even surpass, human-directed grieving. Psychologists have been working to better understand the human experience of pet loss and to address the problem of disenfranchised grief—grief that is not openly acknowledged, socially acceptable, or widely supported—and the complexities of grief complicated by the fact that guardians have so often orchestrated their companion's death through euthanasia.

See also Emotions; Epitaphs; Loyalty; Love

Ground scratching

In the annals of "Why does my dog do that?" you will most certainly find ground scratching. Dogs often scratch the ground after peeing or pooping, dramatically flicking one leg and then the other backward into the air, all the while maintaining a very dignified facial expression. Although nobody knows for sure what dogs are doing when they ground scratch, it is likely a form of social communication, a performative display. One research study found that dogs are more likely to ground scratch if other dogs are around, suggesting that the scratching motion is a visual signal, perhaps of intimidation, although it could mean many other things. The scratching leaves a visual mark in the ground, which dogs who come by later might notice. Ground scratching also spreads and amplifies olfactory messages, both by flinging odors from the anal glands into the air and around on the ground and perhaps also by adding scent from the paw pads. It is possible that ground scratching sends an auditory signal to dogs close by, too. Male and female dogs both ground scratch. Some dog guardians believe that their dog, by scratching, is trying to bury their poop. But this is unlikely, as poop is a gold mine of canine information that really needs to be on display.

See also Communication; Defecation; Olfaction; Urine

Growling

A growl is a low-frequency vocalization, mainly used in agonistic situations to convey a threat. As with a bark, dogs can assess the size of another dog by its growl (a Chihuahua growl sounds different from a rottweiler growl). Growls can also be playful, such as the growling that occurs during a tug-of-war game. Growling is generally higher-pitched in this context. Dogs can differentiate among the various kinds of growls, although having other signals combined with the growl (bared teeth in the context of an agonistic encounter; the play bow during a lighthearted romp) can help dogs read growls accurately. Humans are also reasonably good at discerning the communicative intent of dog growls, particularly those of dogs they know well.

See also Communication; Cur; Play

Guilt

Humans ascribe guilt to dogs with great generosity—more generously, even, than we pass out dog biscuits. Dogs are guilty of all sorts of infractions, including stealing bread off the counter, tracking mud on the carpet, knocking over the trash can, and ignoring us if we say "Come!" when they would rather chase a rabbit. The moralizing of dog behavior is hard to resist, but it is almost always unscientific. How do we humans know that our dog feels guilty? She gets that "hangdog" look, averts her eyes, perhaps lowers her head, or looks to the side. But these are not the behaviors of a guilty dog; these are the behaviors of a dog whose human is expressing anger. Dogs probably do experience guilt, but we don't know enough to say much about it, and the behaviors we typically label "guilty" are most often something else altogether. In one study of dog guilt, cognitive scientist Alexandra Horowitz found that dogs show signs of "guilt" whether they have done something wrong or not, suggesting that it is our expression of disapproval that elicits the hangdog response.

See also Communication; Emotions; Facial expressions

Hackles

"Hackles" refers to hair that sticks up on the back of a dog's neck. Raised hackles can be a sign of fear, and a sign of excitement or surprise. Raised hackles are not, as commonly assumed, a sign of dominance or aggression. The hair follicles on a dog's neck, back, and tail have little piloerector muscles attached to them—called the *arrector pili*. In a relaxed state, the hair follicles rest at about a 30-degree angle to the skin. When activated into fight or flight, the sympathetic nervous system releases the hormone epinephrine. Epinephrine triggers the arrector pili muscles to contract, causing the hairs to stand upright, or "piloerect." This is an involuntary response. A puffier coat makes a dog look bigger and more intimidating. Some dogs raise all

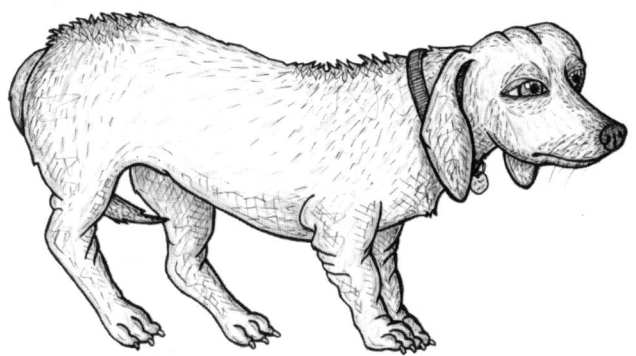

the hair from neck to tail, in one long rib of hair; others just near the neck; others near neck and near tail, rather like a dorsal fin.

Piloerection is also activated by cold temperatures, rather like goose bumps in a human. The sensation of cold activates the sympathetic nervous system, causing contraction of the muscles. When the hairs stand up straight, air gets trapped between hair shafts and creates an insulating layer.

See also Communication

Hayırsız Ada Dog Massacre

In 1910, the governor of Istanbul decided to clean the streets of stray dogs, as part of a larger municipal modernization program. Somewhere between 60,000 and 80,000 dogs were rounded up and shipped off to an island called Sivri Ada, in the Sea of Marmara, where they were left to fend for themselves with no real source of sustenance, other than the cannibalizing of each other's flesh. The dogs died of hunger or thirst or drowned as they tried to escape the island. An archived photo shows hundreds and hundreds of dogs standing together (in a state of disbelief and confusion?) along an island bluff, their fate hanging in the air like a dark ocean fog. A severe earthquake which struck the area in 1912 was perceived by residents as a punishment for abandoning the dogs, leading some to refer to Sivri Ada as Hayırsız Ada ("the inauspicious island"), and the whole event has become popularly known as the Hayırsız Ada Dog Massacre.

What's interesting about this massacre, and any other we might choose, is how profoundly complex the role of dogs in society can be, and how delicately balanced with human cultural, social, political, and religious norms. Scholarly attention to dogs in the Ottoman Empire, and to street dogs in Turkey and in Istanbul in particular, has revealed the multilayered and complex set of circumstances out of which this event unfolded. The Hayırsız Ada massacre is situated within the Young Turk Revolution of 1908 and the introduction of a new ruling party, the Committee of Union and Progress, intent on Westernization and modernization. Stray dogs in Istanbul were associated, by some, with uncivilized culture and social chaos, and modernizing required that they be cleaned out. The removal of dogs was also connected with how the city's garbage was handled; as part of the modernization program, garbage was collected and thrown into large heaps outside the city walls. The work of street dogs in eating garbage had been outsourced, their jobs made obsolete. The residents of Istanbul were not uniformly in favor of the plan to remove street dogs. Indeed, many liked the dogs and had friendly relations with them and found them useful as guardians and as garbage cleaner-uppers.

This canicide is often presented as a one-off, an extraordinary and horrifying event that occurred once somewhere (foreign) and sometime (when we were still brutes). And it certainly has compelling cinematic elements that invite retelling. The island, the earthquake, the cannibalizing dogs. The sheer scale of the horror. But the fact of the matter is that this is just one among hundreds, probably thousands, of such large-scale acts of brutality against dogs. Indeed, mass culling of dogs is completely unremarkable. It has happened countless times and is still happening around the world today, even if in the sanitized and euphemized form of shelter euthanasia. Consider a few modern examples. Prior to the 2014 Winter Olympics in Sochi, Russia, the city hired a private company to round up and kill as many stray dogs as possible. A culling of strays also took place prior to the Summer Olympics in Beijing, China, right on the heels of another culling of roughly 50,000 in Yunnan province, just to get dogs off the streets. The U.S. collects and culls about 400,000 stray dogs each year.

The Hayırsız Ada massacre, in addition to highlighting our propensity toward violence against dogs, is a reminder that we should resist any temptation to make cultural, religious, historic generalizations about dogs and dog-human relationships. For example, from time to time, you might hear someone say that Muslims don't like dogs, that they consider them unclean. Yet this is a gross overgeneralization and captures only one tiny slice of a much larger picture. Beyond that, attitudes toward dogs in Turkey were, and still are, exceedingly complex—as is true everywhere dogs and humans live together. Yes, street dogs are sometimes reviled and murdered. But they are also cherished and loved and considered colorful denizens who belong in the city. Although not likely a direct reaction against Hayırsız Ada, it is now illegal in Turkey to capture or kill stray dogs and has been since 2004.

See also Boji; Cultural attitudes; Diet; Free-roaming; Stray

Head shaking

Why might a pet dog vigorously shake her purple squid toy from side to side, sending stuffing in all directions?

She may be replaying part of a predatory sequence called the kill shake. Head shaking while biting something increases the intensity of a bite and can break the neck of prey. After "killing" the purple prey, she may then move on to evisceration/dissection, carefully opening seams and removing the plastic squeaker along with every last bit of stuffing.

If a dog is repeatedly shaking her head while not holding a toy or dead animal, it could be that she has something in her ear causing it to itch or hurt. Head shaking can be a sign of ear infection, allergies, ear mites, or even neurological disorders, and is a good reason to visit the vet. Head shaking is not the same as a full body "shake-off," which looks like what you might see when a dog has just come out of the water after a swim. Researchers don't really understand the function of the full body shake-off, and research is under way. Dogs seem to do the shake-off in transition times, so perhaps it is a form of stress relief.

See also Ears; Predatory sequence

Head tilting

Dogs cocking their heads at us when we speak is about as iconic and endearing a mannerism as wagging their tails when they greet us. The head tilt has been taken to mean puzzlement or curiosity, or perhaps a dog trying to better hear exactly what we are saying by reorienting her ears. Any of these and more could explain the behavior; we just don't know, because very little scientific research has been done on head tilting.

What little research we do have offers some intriguing clues. Dogs seem to tilt their heads in response to human vocalizations. But why? One research study explored this question by comparing head tilting frequency and direction among a group of dogs, some of whom were considered "typical" and some "gifted." Gifted, in this context, referred to dogs (all of whom happened to be border collies) able to memorize the names of at least ten different toys as taught to them by researchers. The gifted dogs were more likely than their peers to tilt their heads when hearing a command to fetch one of the toys. Because the head tilting took place when a dog was matching the name of a toy to a visual image of the toy, the researchers speculated that it could be a sign of increased attention, and perhaps also that it occurs when a dog is processing known, meaningful stimuli. They also found that some dogs tended to tilt their head to the right, while others usually tilted to the left; there were "right-tilters" and "left-tilters." Given previous evidence for lateralization in the processing of human vocalizations, it may be that the direction of head tilt is lateralized, like pawedness, nostril use, and tail wagging.

Head tilting is to be distinguished from head turning. Tilting, as the word suggests, involves a dog moving her head out of the vertical plane, creating a slope from ear to shoulder, while turning involves rotation on the same vertical plane. Preliminary research on head turning in dogs suggests that dogs most often turn their heads to the left when hearing a spoken command that they recognize.

The few studies conducted on head tilting have looked at the behavior in pet dogs, especially as they interact with their human guardian or researcher in a cognition lab. Nobody has examined whether free-roaming dogs engage in head tilting or under what conditions.

See also Lateralization

Homed

"Homed" is sometimes used as a modifier for "dog," as a more objective and less patronizing alternative to "pet." Homed refers to a dog's lifestyle: Does the dog

identify with and depend for sustenance on a single human dwelling and live mainly inside this dwelling? "Intensively homed" has been used to refer to homed dogs whose movement, social interactions, food options, and behavior are tightly controlled by a human; the dog has no choice about whether and when to be inside the home. "Homed" may be more descriptive and less offensive than "pet," but it is problematic in its own way—as are all terms used to label dogs. "Homed" implies that feral and free-roaming dogs without a human home are homeless, which is inaccurate. These dogs may not have dwellings, but they do have home ranges, and some have places they identify as safe and comfortable and "theirs."

About 20 percent of the world's dogs are categorized as homed, although some of these homed dogs are free-roaming at least some of the time (for example, they may be able to roam their neighborhood), making them hard to fit into just one category.

See also Free-roaming; Nomenclature; Pet; Shelter; Stray

Hound

The word "hound" has its origins in Old English and Middle English, where it was spelled "hund." The etymology of the word can be traced back to Proto-Germanic, the ancestral language from which many modern Germanic languages, including English, evolved. In Proto-Germanic, the word *hundaz* referred to a domesticated canine or dog. *Hund* still has that generic meaning in modern German, but in English, it was upstaged by "dog," a word of untraceable origin, while over time the term "hound" has come to be associated with certain breeds that possess specific characteristics suited for tracking scents, chasing prey, and assisting hunters. While the word's basic meaning has remained consistent, the specific breeds referred to as hounds, and their uses, have diversified. Also, as slang, "hound" has been used to refer to a man who goes after women in an exhausting, irrepressible fashion.

See also Cur; Dog; Mutt

Humping

Humping is a rhythmic canine thrusting of the hips reminiscent of John Travolta in *Saturday Night Fever*. Humping often occurs after mounting, though sometimes humping occurs without mounting. This "air humping" looks suspiciously like a dog trying to twerk. Male and female dogs both hump, will hump dogs of either sex, and will hump whether intact or altered (neutered/spayed). Dogs will also often hump things that aren't dogs, including human legs and arms, blankets, dog beds, pillows, and whatever else is readily available and roughly the right shape and size and texture.

In male dogs, humping often leads to an erection, even if the humping is nonsexual. "Arousal" is a state of heightened physiological excitement and is not necessarily sexual in nature. Arousal can lead to penile erection. Humping can also be a displacement behavior; some dogs will hump in response to a stressful or exciting event. Like mounting, humping is also part of the repertoire of play behaviors.

See also Displacement behavior; Mating; Mounting; Play; Red Rocket

Hybrids

A hybrid is the offspring of two animals of different species. (Mixed-breed dogs are not hybrids.) Naturally occurring hybridization has been observed between dogs and wolves, coyotes, dingoes, and jackals, in areas where there is substantial overlap in home ranges. For instance, a 2014 study in the Caucasus region between the Black Sea and the Caspian Sea found hybrid activity in every tenth wolf and every tenth shepherd dog. Humans also intentionally breed hybrids, out of a desire to create an interesting and exotic new kind of pet. Wolfdogs and coydogs have been two of the most popular hybrids. It is generally thought that these animals have poor welfare because they are behaviorally unsuited to captivity and human friendships.

An interesting example of a hybrid is the Sulimov dog. The Sulimov is named after the Russian mad scientist Klim Sulimov who, in the 1970s, decided to create a Superdog with supernatural smelling skills, to be used for airport security by Aeroflot, the largest Russian airline company. Sulimov reasoned that the domestication process might have weakened dogs' olfactory sense, because the dog no longer must hunt its own food; wild canids, on the other hand, have retained their sharp noses. And the canid species with perhaps the most superb sense of smell is the jackal. Sulimov also believed, mistakenly, that the jackal was the closest evolutionary kin to dogs, and he envisioned taking dogs back to their original, undiluted (by human selection pressures) state. With these beliefs in mind, Sulimov bred dogs and jackals together, in varying iterations, to get what he considered the ideal balance of trainability and olfactory smarts. The result, the Sulimov dog, is a dog-jackal hybrid (¾ dog, ¼ jackal) bred specifically for its sense of smell. The jackal-dog breeding program is still in operation, and these hybrids are still working at Moscow's main airport, as narcotics detection dogs. It is worth noting that although Sulimov swears by his dogs, scientific evidence that they have superior sniffers is weak at best.

The Sulimov raises interesting questions—not well understood—about whether certain breeds of dog have better olfactory skills than others and, if so, whether this relates to genetics, to environmental factors such as training, or to both. (The answer is almost certainly both.) Scientists studying how to maximize the olfactory skills of dogs have, like Sulimov, tried hybridization; they have also tried selective breeding, as well as some experimental manipulations that belong in the Horror Box, such as sewing shut the eyes of puppies to see if they would compensate for their blindness by developing a keener sense of smell (they didn't).

See also Breed; Mixed-breeds; Landrace; Olfaction; Working

Inbreeding

Inbreeding refers to mating between genetically related individuals, like mating with your sibling or, as popularized by European monarchies, with close cousins. Inbreeding is rare in nature, unless there is a genetic bottleneck caused by ecological conditions such as isolation of a species to a small patch of habitat. In natural populations there are mechanisms in place to prevent inbreeding between genetically related individuals, such as dispersal, olfactory cues signaling "don't breed with someone who smells like you," or pack hierarchies based on age and genetic relationship.

In contrast, inbreeding is a significant issue for pure-bred dogs, according to studies published over the last few years. Indeed, most "officially" recognized dog breeds are highly inbred. Many breeds have a small founding population and are then consistently bred together to "fix" (or lock in) certain traits, usually related to physical appearance, leading over time to a loss of genetic diversity. Inbreeding increases the chances that offspring will carry genes for recessive traits, which tend to be those that reduce biological fitness. This is why highly inbred dog breeds can have shorter lifespans and are at greater risk for certain cancers, orthopedic disabilities, and other diseases.

See also Artificial selection; Purebred; Selective breeding

Inequity aversion

Although moral behaviors were long assumed to be the special province of humans, research over the past several decades has blown open the doors. A whole range of behaviors which in humans we label "morality," including empathy, cooperation, honesty, trust, and a sense of fairness, have been observed in a wide range of species, suggesting that moral behaviors have evolved broadly to help social animals get along together in groups.

Behaviors related to a sense of fairness have been of particular interest to scientists. Within a group of animals, distribution of resources needs to be relatively equitable for cooperation to be evolutionarily beneficial. So, what mechanisms are in place to maintain fairness in distribution? One mechanism is a negative response—a "Hey! That's not fair!"—to the unequal distribution of resources ("rewards") or to inequitable outcomes. Scientists call this "inequity aversion" (IA). Researchers have come up with several behavioral scenarios that might suggest that an animal is feeling the burn of unfair treatment: an individual might refuse to work under unfair conditions or might reject a reward that is considered unfair. One common way to test for IA is a so-called token exchange. In the experimental setup, two conspecifics give tokens to an experimenter in exchange for rewards of varying qualities. If one animal observes another getting a better reward for the same token, does this elicit screams and fits of anger or just a shrug of the shoulders? If it elicits anger and noncooperation, the animal is likely experiencing IA.

Dogs have been asked to do such task-for-reward exchanges. In one early study, researchers asked dogs to give their paw in return for a treat, in the presence of other dogs doing the same task. If another dog got a better treat, it didn't seem to be upsetting. But if one dog got a treat and the other got nothing, the shortchanged dog would refuse to respond to the "paw" command. In other words, dogs don't seem to respond negatively to inequity in the quality of reward but do seem to respond negatively if a partner is rewarded and they are not. Although various studies have explored IA in dogs, there are many questions that researchers are still trying to answer.

See also Emotions; Play

J acobson's organ

The domestic dog has two major mechanisms for detecting odor, the main olfactory system and the *vomeronasal organ* (VNO), which work as a complex, integrated whole. The VNO is often referred to as the Jacobson's organ, named after Dutch army surgeon and sometime anatomist Ludvig Jacobson, who first identified the organ—which is present in many mammals, reptiles, and amphibians—in the early nineteenth century.

The VNO is a paired right and left set of organs that sit on the underside of what's called the vestibule, inside the nasal cavity, and which open into the roof of the mouth behind the upper incisors. In simplified terms, the VNO connects the nasal cavity with the mouth cavity. The VNO is the main structure involved in recognition of chemical signals called pheromones and other low-volatile substances, and may play a role in taste, although the precise function of the organ is not fully understood.

The nerves from the VNO pass not to the main olfactory bulb in the brain, but to an accessory olfactory bulb, which has connections to regions of the brain that control social and sexual behavior. When a dog sticks his tongue into another dog's pee spot, he is likely using the VNO to smell/taste pheromones in the pee. You might also notice, during the pee smelling/tasting, that the dog's front teeth are chattering. This is called the *flehmen* response: the chattering facilitates the movement of chemical particles through Jacobson's organ.

See also Communication; Noses; Olfaction; Sniffing

Jealousy

Do dogs experience jealousy? Most people who live with a dog will answer with a resounding yes and will have a whole collection of stories about what makes their dog green with envy and what sorts of behaviors jealousy seems to provoke. Indeed, the term "dog-cheating"— when you come home with the scent of another dog on you—has a firm place in the modern lexicon of pet-dog keeping. Science seems to confirm dog guardian intuition, though much work remains to be done before scientists have a thorough understanding of canine jealousy.

What does the science say? There is accumulating evidence that dogs do experience jealousy, but we still know relatively little about how, why, and when they do, and ascribing the emotion "jealousy" to a certain behavior is inexact and fraught with complications. The emotion of jealousy may have evolved in a wide range of social species, as a mechanism to protect social bonds from the intrusion of third-party interlopers (informally known as "cuckoldry"). The functional role of jealousy may be protection against the fitness consequences of such losses (e.g., having someone else's genes, rather than your own, carried forward to future generations), which has expanded behaviorally to encompass the protection of emotional bonds in a wide range of valued relationships, including friendships.

Although scientists seem generally to agree that dogs experience a social emotion akin to jealousy, what has not yet been sussed out are the nuances among jealousy, envy, and inequity aversion, and what emotional state underlies putative jealous behaviors. Jealousy in dogs may be a form of social resource guarding, a feeling of being threatened by the potential loss of one's social attachment. Envy, in contrast, may be the feeling of wanting what someone else has, especially possessions such as juicy bones or giant bowls of food. Inequity aversion, for its part, is a negative response to the perception of being treated unfairly, and not getting the same reward as someone else who has made a similar contribution.

What do jealous behaviors look like? Some possibilities include body blocking, snapping, barking, and pushing against an interloper with nose or body. Like people, some dogs may be more prone to jealous behavior than others.

See also Emotions; Guilt; Inequity aversion

Joy

Do dogs experience joy? Without a doubt. Joy is visibly embodied in a dog's movements. Dogs will sometimes "dance" with joy, assuming a large, rather stiff posture and bouncing on their front paws. This dance of joy is often observed in response to a dog's human returning home from an absence and in anticipation of a walk or hike—the dog bouncing around as the human (too slowly) puts on shoes and coat and searches for the leash. Sometimes dogs bark as they do this joy dance. When else might dogs feel joy? Play bouts with other dogs seems like an obvious candidate. Dopamine is released during play, and likely also serotonin and norepinephrine, all of which are associated with feeling good. Researchers have identified a specialized vocalization, that canine form of "laughter" called the play-pant, which occurs during human-dog play. Perhaps this laughter is associated with joy—we don't know for sure. "Zoomies" might also be an outward expression of joy.

A study from 2022 suggested that dogs, like humans, may shed "tears of joy" when experiencing strong positive arousal. Researchers found that tear volume increased in dogs when they were reunited with their owners after an absence. Tear secretion is mediated by the hormone oxytocin, which plays an important role in social bonding. It may be that these doggie tears have evolved because they elicit an emotional response in humans, making us want to engage in caregiving behavior.

Despite its obvious appeal, little official scientific research has been conducted on dog joy. Some open questions about canine joy: What is the difference between canine happiness and canine joy? Is the difference quantitative or qualitative? How and when, exactly, do dogs experience joy? How is dog joy like and different from the joy we humans experience? What is the role of joy in the canine good life? These are all questions of interest.

See also Emotions; Zoomies

K-9

The use of K9 or K-9, a homophone of "canine," to refer to army and police dogs may have originated during the Second World War. The first U.S. military K-9 Corps was approved in 1942, by Under Secretary of War Robert P. Patterson. "K-9" was registered by the U.S. Army in 1959 as a federal symbol in the U.S. Patent and Trademark office: "The term 'K-9,' as applied to dogs and dog equipment, is associated with the Quartermaster Corps." The use of "K-9" has broadened but is still especially associated with military and police dog units. Police dog units have the "K-9" symbol on their vehicles, and military dogs wear vests inscribed with "K-9."

Dogs have been used in law enforcement since at least the Middle Ages. (Then as now, there has been a thin line between the use of dogs as a tool of law enforcement and as a tool for subjugation and exploitation of certain groups of people.) Formal training of dogs in law enforcement goes back over 200 years. Police dogs in the U.S. today are used in patrol (including crowd control during protests, protecting officers from harm, apprehending suspects, tracking missing persons) and detection (drug and bomb sniffing). The breeds most often used in policing, as in the past, are large, powerful, and "biddable" (trainable), such as German shepherds and Belgian Malinois. Breeds known for their exceptional sniffing skills are chosen for tracking work: bloodhounds, basset hounds, Labrador retrievers.

K-9 is distinct from Animal Control, which is a branch of law enforcement whose function is to use people to control dogs, not vice versa. Animal Control primarily enforces laws which prohibit dogs from roaming free. Some of the tools of Animal Control include catchpoles—long poles with loops of wire at the end, used to put physical distance between an officer and a potentially aggressive dog—and special trucks outfitted with cages for impounded dogs.

See also Racism; War; Weaponizing; Working

Kibble

Although for nearly our entire history together dogs have eaten what humans have eaten, in 1860 a strange idea took hold: dogs should have their own separate food, which we will manufacture in huge factories, put into large bags, sell for lots of money, and which will consist of extruded and heat-processed bits of hard material that we'll call "kibble." From its humble beginnings, kibble has become part and parcel of pet-keeping practices. Pet dogs are tightly restricted from eating their natural diet of foraged and scavenged items; canine efforts to access scraps of human food are labeled as naughty behavior; an army of behavioral supervisors called Trainers are employed to make sure dogs only eat "dog food."

The invention of dog food was one of the greatest entrepreneurial successes of all time: an entire new category of need was invented and then filled. The first commercially prepared dog food, and our earliest iteration of kibble, was introduced by businessman and electrician James Spratt. As he watched dogs being fed leftover hardtack—what sailors ate during long ship voyages—he had the idea that dog owners, too, could use a form of shelf-stable food for their pets. Spratt's original dog biscuit was made from the "dried unsalted gelatinous parts" of cows (perhaps cow blood?), beetroot, wheat meals, and vegetables.

Many homed dogs are fed some sort of manufactured dog food, either dry kibble which comes in a bag or wet food that comes in a can. Like Spratt's biscuit formulation, these foods generally contain a protein source—often the rendered parts of slaughtered animals that humans find repulsive, such as the hooves,

beaks, noses, and hair—combined with grains, vegetables, or fruits, as well as additives that make the smell
less offensive to humans and preservatives that stabilize
the food for a long shelf life. Although kibble is con-
venient, some concern has been expressed about the
highly processed nature of the food. A dog subsisting
on ultra-processed dog food is rather like a human subsisting on fast food burgers and fries.

See also Diet; Food resources; Pet

Laika

On November 3, 1957, a small husky-spitz mix
who grew up on the streets of Moscow made big
history. Nicknamed Laika ("Barker") by Soviet scientists who trained her, she was launched into orbit on
Sputnik 2 and became the first living creature to orbit
the Earth. *Sputnik* 2 exploded into a ball of fire as it re-
entered the Earth's atmosphere, but by this time, Laika
had already perished inside the capsule.

The quest for the perfect canine candidate began with gathering a whole group of strays from among the many dogs living on Moscow's streets. Female dogs were chosen because they were smaller and supposedly more tractable than male dogs. The candidate dogs were subjected to various exertions, such as being confined, first for days and then for weeks, to tiny, pressurized capsules. Two finalists were chosen: Kudryavka (a.k.a. Laika) and Albina. Both dogs had medical devices surgically implanted that would monitor heart rate, blood pressure, and other vitals. Laika was ultimately chosen for the mission, in large part because one of the top scientists involved with the project had developed a special affection for Albina and didn't want her to suffer.

After takeoff, Laika lived for 103 minutes. Scientists had expected Laika to die from oxygen deprivation after her limited supply ran out, but instead she died of heat stress, as the loss of the heat shield caused temperatures inside the capsule to surge. There was never any intention to bring Laika back alive; the technology to return the capsule safely to Earth did not exist. Although Laika was the first dog to orbit the Earth, she was neither the first nor the last dog to be sent into space. Indeed, the Soviet space program would not have been as remarkable as it was without the contribution, however involuntary, of many dogs. The Soviets' choice of dogs rather than primates, which were primarily used by the American space program, may be what gave them an edge.

"Laikas" are also a type of landrace dog (see next entry) found in northern Russia and parts of Europe, employed as hunting partners and perhaps also as sled pullers. Exactly which dogs belong in the category of laika dogs is disputed. Several different laika breeds are recognized by the American Kennel Club and the Fédération Cynologique Internationale.

See also Epitaphs; Landrace; Working

Landrace

A landrace is a group of genetically related dogs unique to a specific geographical location and often serving a function in local human agricultural practices, such as livestock herding, livestock guarding, or sled pulling. Landraces are well adapted to local environmental conditions, including altitude, temperature, terrain, and abundance or scarcity of water. Although landraces share a certain level of physical uniformity, they don't conform to breed standards or belong to breed clubs and registries. Landraces have been bred for survival in a particular place, with greater attention to function and hardiness than visual appearance. The terms "landrace" and "ancient" or "primitive breed" are sometimes used interchangeably. A few examples of landrace dogs: Carolina dog, saluki, Ibizan hound, Xoloitzcuintli, Anatolian shepherd, and Maremma sheepdog.

Landrace dogs sometimes have special adaptations (genetic changes) that have helped them flourish in their home environment. Huskies, for example, have adaptations that help them thrive in cold temperatures. Tibetan mastiffs, who were bred as flock guardians on the Tibetan Plateau in the Himalayas, are particularly well-adapted to their high-altitude ecological niche. They are also an interesting example of *heterosis*, or interbreeding with benefits. The Tibetan mastiffs may have interbred with Tibetan gray wolves, who are hypoxia tolerant; this may explain the adaptive mutations to hemoglobin seen in these mastiffs.

See also Breed; Domestication; Hybrids; Laika; Purebred; Xolo dog; Zhokhov Island

Lateralization

In all mammal species and most vertebrates, the left and right hemispheres of the brain specialize in different kinds of cognitive processing. This functional specialization of the brain is referred to as *lateralization*. Lateralization in dogs has been studied in relation to paw preference, tail wagging, and nostril use, among other things.

Handedness is one of the most common measures of laterality, both in humans and in dogs. Paw preference, or pawedness, has been studied in a variety of ways, two of the most common (and least welfare-compromising) of which are the Paw Task and the Reach Task. In the Paw Task, a dog in a sitting position is presented with a guardian's or researcher's outstretched hand, as if asking a dog for a "high five." Whichever paw the dog lifts in response is considered dominant. In the Reach Task, an object the dog is likely to desire, such as a treat or a toy, is placed just out of reach, and situated so that a dog will need to use her paw and not just her mouth to retrieve the object. Again, whichever paw the dog uses for the task is considered dominant. Right-pawedness is more common in dogs, as in humans; also as in humans, there is a sex bias, with a greater percentage of females showing right-pawedness (and right-handedness). A recent study tried to figure out if the handedness of a dog's guardian influences the paw preference of the

dog. Although the study was small—only 62 dogs—
the results are intriguing: dogs living with left-handed
humans were more likely to show left-paw bias, while
dogs living with right-handed owners were more likely
to display right-paw bias, suggesting that some degree
of learning is involved in paw preference.

Dogs also show lateralization in tail wagging: when
wagging, their tails move more to one side than the
other. Watch some dog tails and see if you can observe
this. One research study observed dog tails while dogs
looked at stimuli that had positive emotional valence
(their human guardian) and negative emotional valence
(a stranger). When dogs looked at things that made
them happy, their wagging was biased toward the right
and when they looked at things that made them nervous
or stressed, they wagged more to the left. Dogs appear
able to use information gleaned from the tail wagging
of other dogs to deduce emotional states, allowing them
to make nuanced assessments of how other dogs are
feeling and to alter their behavior accordingly.

Dogs have strong right-nostril bias; they start a sniff
with the right nostril. If the smell is familiar and not so
interesting, the dog shifts to the left nostril. If, on the
other hand, the smell is new, potentially threatening,
or especially arousing, they stick with the right nostril.
In one study, dogs were presented with arousing odors
like the smell of food, the sweat of the dog's vet (ew!),
or vaginal secretions of a female dog in estrus. The
dogs preferentially used the right nostril when smelling physiologically arousing stimuli. Since, unlike all
other senses, olfactory organs "report" to their same-
side brain hemisphere rather than cross over, scientists

have hypothesized that the right hemisphere of the canine brain controls processing of novel information and regulates the hypothalamic-pituitary axis (the fight-or-flight response); the left hemisphere controls behavioral processing of familiar stimuli. The study of lateralization has also filled in some details about how dogs use chemical signals to communicate, including how they might use odors to "read" the emotional states of other dogs and, also, of people. One small study found that dogs preferentially used their right nostril to sniff odors from another dog who was emotionally aroused (from being isolated) but used the left nostril to sniff odors from a human who was emotionally aroused (from watching a scary movie).

See also Head tilting; Paws; Tails; Wagging

Leashes

Leashes have been around for millennia. An Egyptian cave painting from 3500 BCE shows a dog tethered to a leash. The leash has for thousands of years been a tool used by humans to control dogs under specific working conditions, such as hunting or guarding. Leashes have also been integral to the practice of pet-dog keeping, especially in modern times. In the late 1800s and early 1900s, self-appointed dog care experts and humane organizations such as the American Society for the Prevention of Cruelty to Animals began to warn owners that letting dogs roam was dangerous, both for the dog and for the dog's owner, particularly because rabies was widespread among dogs at the time. Roaming pet dogs might get rabies from a stray, and then pass the disease to their owner's family. The advent of motor cars added to the safety concerns for dogs. Leash ordinances began appearing in the late 1800s in large cities like London, Paris, and New York. Leashing is now fully cemented as a normalized practice and a necessary component of responsible pet ownership.

The purpose of a leash, of course, is to control the movement and behavior of a particular dog and, by extension, all dogs. In some countries, such as the United States, municipalities often have strict "leash laws" that require all dogs to be leashed when not on their owner's property or in a specially designated off-leash area. Being unleashed leads to impoundment. In this way, leashes reinforce the cultural categories of "pet" and "stray."

Leashes can profoundly mediate how dogs experience and engage with the world around them, mostly in ways that are unfortunate for dogs. Leashes obviously constrain dogs' movement, influence where they will be able to scent mark, what they will be able to smell and see, the pace, direction, and speed of their explorations. The fact that there is an entire category of behavioral disorder called "leash reactivity"—canine behaviors, such as lunging at other dogs or human pedestrians, that occur in reaction to being leashed—points to the profound welfare implications of this seemingly simple tool of control.

See also Collars; Cursorial; Pet; Stray

Lenticular sclerosis

Lenticular sclerosis is the medical term for the cloudy, sometimes bluish or greyish haze that develops in the eyes of nearly all our cherished canine elders. It can begin during middle age (6–7 years of age) and is considered a normal age-related change. The lens fibers thicken and harden, but sight is not significantly impaired. Dogs can also get cataracts, which do cause diminished vision. Cataracts can be treated surgically in dogs, just as they can in humans.

See also Tapetum lucidum

Licking

Licking is a behavior with many uses. Dogs lick themselves as a form of grooming, wound care, or self-soothing; they lick each other in greeting, as a sign of affection, and as a calming signal. Dog mothers lick puppies to clean them; puppies lick their parents' mouths to prompt regurgitation and feeding. Being licked releases oxytocin, a hormone associated with social bonding, in the dog on the receiving end; it also releases oxytocin in the dog doing the licking.

Evidence suggests that licked wounds in dogs heal twice as fast as unlicked wounds. Nevertheless, dogs can overdo it and can increase inflammation by too much licking, which is when the dreaded "Elizabethan collar," a.k.a. "the Cone of Shame," is pulled out of the cupboard. Compulsive licking can also be a sign of psychological distress and is considered a behavioral pathology. Excessive licking, sometimes caused by allergies and sometimes caused by psychological distress, can lead to what is called a lick granuloma, or acral lick dermatitis, an inflamed skin lesion which itches and hurts and invites more licking, leading to a vicious cycle that requires intervention.

We must also comment on the phenomenon of dogs licking humans, which many dog guardians optimistically refer to as "doggy kisses." These slobbery kisses could be a sign of affection; licking may also be an experiment to see what the human had for lunch and an excavation to see if any interesting flavors remain on the face.

Dogs also lick people's wounds. For a long time, people believed that being licked by a dog would facilitate the healing of wounds. In the Temple of Asclepius, the ancient Greeks employed sacred dogs as wound lickers for sick petitioners. This may sound bizarre to modern ears, but dog saliva may, in fact, have some healing power. Dog saliva contains several different proteins with antimicrobial properties, including histatins, simple proteins that defend against infection. On the cautionary side, however, dog saliva can be infected with oral bacteria and viruses and can spread disease. The composition of dog saliva can be altered by diet, age, environmental factors, health condition. And "compromised" saliva may increase the risk of *zoonosis*, or animal-to-human disease transmission. Dog saliva can carry bacteria that cause salmonellosis, leptospirosis, and campylobacteriosis, among other things.

Also, if you are free and loose with face licking, you might bear in mind that dogs often lick their heinies after they poop—they like to keep themselves clean! And, really, dogs lick "down there" at most any time of the day or night. Dogs often greet one another by sniffing butts. So, some parasites and bacteria found in dog feces could potentially wind up in their saliva. Dogs also eat stuff that is gross, like feces and dead things and trash. So, yeah, just remember that.

See also Affiliative behavior; Parasites; Saliva

Literature

Dogs have been part of the literary landscape for almost as long as people have been writing. Consider the titles of two ancient tablet-inscribed narratives written about 2000 BCE, neither of which will likely be included in dogs' politically acceptable library and which might even be purged from their canon: "The Show Dog" and "Why the Dog is Subservient to Man."

Human literature would certainly not be what it is without dogs. Dogs have served as metaphors, both for depicting human strengths and for highlighting our character flaws. Dogs have played countless bit parts in our stories. And dogs have been at the very heart of some of the most beloved books in the Western canon. Some of these classics have unforgettable dog characters (Steinbeck's *Travels with Charley*), some are built around unforgettable human-dog friendships (*Where the Red Fern Grows*, *My Dog Skip*, *Old Yeller*). And in some, such as *Call of the Wild*, an unforgettable dog is the book's protagonist, with humans playing supporting roles.

See also Argos; Balto; Loyalty; Movie stars

Love

Perhaps it would be an oversight not to include an entry on love, the heart and soul of the human-dog bond. We say dogs are the best possible friends and family because they love us unconditionally, unlike our human friends and family, who seem so often to have hidden agendas and who judge us for our shortcomings. But do dogs really love us? Surely—but perhaps not as unconditionally as we'd like to think. They may follow the Christian dictum and love our neighbors as much as they love us, if the neighbors are nice and give good treats.

Is love the reason dogs and humans live together? No. It is far more complicated than that. Do we love our dogs? We say we do, but our behavior isn't always consistent with our pledge.

Less often noted in conversations about dogs and love is that dogs form romantic attachments with each other, a fact that we often seem to look past in our aggressive manipulations of canine reproduction. Dogs who live as pets rarely get to form love relationships, and when they do, they rarely get to consummate their love. Like humans, dogs should be able to love who they love. Dog mothers also love their children, abundantly, a bond often and painfully fractured by the human practice of taking young puppies and making them our own.

See also Epitaphs; Grief; Loyalty; Mating

Loyalty

Among the moral virtues assigned to dogs, loyalty is perhaps the most common. To pick just a few from the massive pile of loyal dogs in our cultural and historical collection, we have the loyal Argos, mentioned earlier, who alone recognized Odysseus in his disguise of rags when he returned from war. We have Lassie, the television star who risked her life over and over and over and over to save her boy. And we have Hachiko, an Akita Inu made famous for his dedication to his human guardian. Hachiko traveled from home every day to the Shibuya train station in Tokyo, Japan, to meet his human as he returned from work. When the man died, Hachiko continued to go to the train station at the appointed time, *for the next nine years*, until Hachiko's own death.

Dogs may, in fact, possess the virtue of loyalty. However, there is no actual scientific evidence for dog loyalty, which is odd given how often we claim this behavior on their behalf.

Some research suggests that dogs prefer their owners to other humans, and even, if you can believe it, that some dogs are more attracted to the odor of their special human than to the odor of food (at least at the time of testing). Another study, which some have taken to bear slightly on the question of canine loyalty: dogs witnessing a stranger being mean to their owners were less likely to take treats from the mean person than from a stranger who hadn't been mean to their owner. Studies of this sort are fun, but don't really tell us much about canine loyalty or lack thereof. What they might tell us, though, is something about social cooperative behaviors. Dogs make social evaluations of people even when not in their direct self-interest—if purely self-interested, the dogs would likely have taken treats from anyone and everyone.

"Loyalty," then, depends on context. Loyalty shouldn't be confused with affection, attachment, dependence, or self-preservation. It may boil down to how we use the word and how we label. Was Hachiko loyal? Or was he simply following deeply grooved habit pathways? Only Hachiko knows the answer to this question, and unfortunately, he passed away over 100 years ago.

Is your dog loyal? Believe what you want.

See also Argos; Love; Man's Best Friend

Man's Best Friend

"The dog is man's best friend." That is undoubtedly one of the most common, and most irritating, sayings in the English language. Why might this phrase get under the skin? Most obviously, dogs are friends with women, too. But also, it is a treacly cliché that overlooks an important fact: dogs aren't necessarily our best friends. They may tolerate us because they have no choice, or because we provide pretty good vittles and a warm place to sleep. Instead of best friends, this might make them artificial or paid friends, or friends with benefits. The "Man's Best Friend" bromide also reinforces the notion that dogs will love us and stand by us no matter how poorly we treat them—which seems to be taken as permission to treat them "like dogs" (an idiom meaning "to treat someone like crap").

An anecdote frequently mentioned in relation to the use of the phrase was an 1870 trial in Warrensburg, Missouri, over the wrongful shooting of a coon hound named Old Drum. Old Drum was allegedly shot for trespassing on a neighbor's property; Old Drum's owner, Charles Burden, sued for damages. Lawyer George Graham Vest's closing argument included a moving eulogy to Old Drum and to all dogs: "The one absolute, unselfish friend that man can have in this selfish world—the one that never deserts him and the one that never proves ungrateful or treacherous—is his dog." *Burden v. Hornsby* went all the way to the Supreme Court of Missouri. Burden was awarded $50. The trial is noteworthy as a very early example of legal suits claiming damages for the death of a dog and addressing the vexing question of how to place monetary value on the life of a beloved canine companion.

See also Epitaphs; Loyalty

Mating

Mating encompasses a whole range of behaviors related to reproduction, including mate choice, the flirtatious signaling that indicates an interest in mating, courtship, competition for mates, copulation, post-copulatory behaviors meant to ensure paternity, and how animals protect and raise young.

Domestic dog mating systems have been under human-influenced selective pressures and differ in interesting ways from the mating systems of wolves. Unlike wolves, who breed in spring and have one reproductive cycle a year, domestic dogs can breed at any time of the year and typically have two breeding cycles. During the breeding cycle, female dogs enter into periods of sexual receptivity, or "heat," which can last a week or ten days. Females in estrus emit pheromones, particularly in their pee, which they may distribute more frequently while in heat by peeing here and there. The female will show increased interest in male dogs and may seek out a preferred male companion. Male dogs, for their part, are promiscuous and will try to mate with as many females as will have them. When a female dog is in heat, male dogs may threaten or fight other male dogs for access.

Behavior that builds up to mating is called courtship, and may include play, licking each other's faces, and otherwise spending time together. A period of courtship allows females to be choosy about their mates, giving

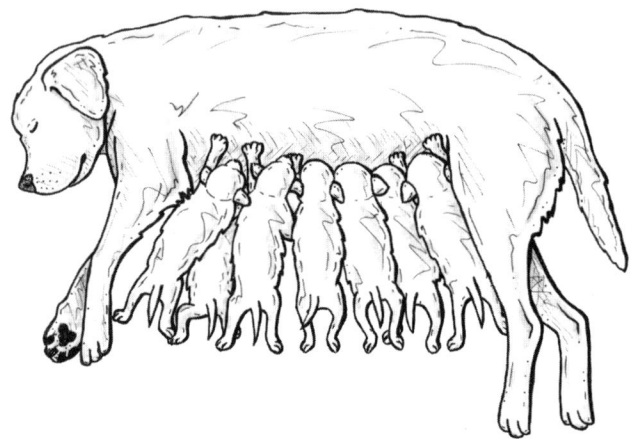

them time to get to know a mate before committing to kids. The courtship and mating period for domestic dogs is often brief, especially compared to wolves. Once a female has chosen a mate, the two will get down to business. If all goes well, a litter of pups will be born 63 days later. A female may mate with multiple males, and more than one male can provide genetic material for a litter of puppies. In a litter of five puppies, for example, each pup might have a different father.

See also Red Rocket; Reproduction

Mixed-breeds

"Mixed-breed" typically refers to a dog whose parentage is unknown or whose genetic background is an admixture of three or more different breeds. A crossbreed, in contrast, is generally understood to be a deliberate human breeding of two purebred parents of different breeds (so, the goldendoodle is a crossbreed of golden retriever and poodle). The monetized version of a cross-breed is a so-called designer dog. Designer breeds are often given a cute portmanteau rebranding, such as pugalier (pug + Cavalier King Charles spaniel), shih-poo (shih tzu + poodle), and Morkie (Yorkshire terrier + Maltese).

People tend to think of mixed-breed dogs as being exactly that: a whole bunch of random purebred bits of DNA, all covered in fur. Purebred + purebred + pure-bred = mixed. A+B+C=ABC. But the algebra may be more accurately seen as A+B+C=D. Mixed-breed dogs are not just an amalgam, an averaging out or adding together of breed types. Instead, as researchers studying dog behavior have emphasized, mixed-breed dogs are a unique group, with characteristic behavioral traits.

Mixed-breed dogs have been neglected in canine re-search, even though they make up most of the world's dog population. Most of the research on dog behavior and health (e.g., Scott and Fuller's genetic studies) has been conducted on purebred dogs, with specific breeds being used as a study group. This is useful because re-searchers can control for obvious breed differences. But there are also limitations with breed-focused research. Small studies on a particular breed are often then gen-eralized into "dogs do this" and "dogs behave like this." This conflation of a single breed of dog with The Dog may divert attention from the idiosyncrasies of dis-tinct breeds and of distinct individuals within breeds. A research landscape so strongly biased toward breeds ignores the potentially profound differences between purebred dogs and mixed-breed dogs, as well as the differences among dogs who live as pets, those who live in research laboratories or kennels, and those who live partly or completely on their own.

One of the questions on people's minds—and a source of many opinionated diatribes—is whether pure-bred dogs are "better" than mixed-breed dogs. But better or worse by what standard? Sometimes these comparisons are based on perceived behavioral differences. People often claim, for example, that mixed-breed dogs—especially those who come from shelters—are more behaviorally challenged, more likely to be aggressive, more unpredictable, than purebred dogs. But the behavioral research is, well . . . it's highly mixed. Some studies have shown mixed-breed dogs to be less obedient, more aggressive, more excitable than their purebred peers. Other studies have shown mixed-breed dogs to be more trainable, less reactive, and so forth. What it boils down to is that reported differences likely relate less to lineage than to other factors such as human expectations, husbandry practices, socialization, and early experiences of the dogs.

Sometimes comparisons are made based on perceived health differences. The most common piece of folk wisdom doled out is that mixed-breeds are healthier than purebreds, on average; have fewer inherited diseases; and live longer, because of their greater genetic heterozygosity—their "hybrid vigor." But this broad generalization simply doesn't have firm empirical support. What we can say, with reasonable evidence behind it, is that many purebred dogs have health and behavioral problems related to their breeding, whether to their being bred for morphological features that we find appealing but that are linked with disability (such as extreme brachycephaly) or because of excessive inbreeding (too little genetic heterozygosity) and the retention of recessive traits that are maladaptive.

See also Breed; Communication; Hybrids; Inbreeding; Purebred

Mounting

Mounting behavior involves climbing partway on top of a conspecific, either from the side or from the back. Mounting often occurs just before humping. Dogs mount each other in several different contexts, including during courtship and mating, during play, and as a sign of dominance. Dogs sometimes mount and then hump inanimate objects. Humans have been observed doing this, too.

See also Humping; Mating; Play

Movie stars

Although no comparative studies have been conducted, it seems safe to say that dogs have appeared in more movies and television shows than any other species of animal, with the possible exception of horses. And dogs have certainly enjoyed the lion's share of starring roles. Unlike human movie stars, who are known by their real names, we usually know canine actors only by their screen names: Rin Tin Tin (apparently there were 14 German shepherd replacements after the original Rin Tin Tin), Lassie (played, in the television series, by 9 different collies, all of them male), Strongheart (the screenname for Etzel von Oeringen), Benji (the original Benji was named Higgins and was adopted from the Burbank Animal Shelter), Sounder, Red Dog, Old Yeller, and so on.

The use of dogs in movies and other visual entertainment has not been without controversy. In the past, and even too often still, dogs are trained for movie roles using punitive methods, or are placed into situations—like having to swim across a storm-swollen river—that are scary or stressful. (Try watching the original 1935 *Call of the Wild* without cringing.) Animal welfare standards for movie and television production have improved, and you will now often see disclaimers such as "No animal was harmed in the filming of this movie."

The number of animated features starring dogs is even more expansive. Many of these are adaptations of beloved books or cartoons, including Clifford, Scooby Doo, and Snoopy. Each, in its own way, contains interesting messages about how we view dogs and human-dog relationships, and the roles dogs play in society. The classic Walt Disney movie *Lady and the Tramp*, for instance, plays out a familiar human trope: guy from the wrong side of the tracks falls in love with girl from the right side of the tracks. Interestingly, Tramp, as his name (or lack thereof) suggests, is a mongrel, while Lady is a pedigreed cocker spaniel. The love relationship between them is socially inappropriate; to be together, they must transgress class boundaries.

Researchers have identified patterns in dog acquisition behavior in which the release of a popular movie starring a dog will lead to notable spikes and dips in the popularity of that breed of dog, the so-called "Hollywood effect." *Rin Tin Tin* raised the profile of the German shepherd; *Strongheart* pushed it even higher. German shepherds are now one of the most recognizable and beloved breeds of dog. After *Lassie*, there was a surge of interest in collies. *One Hundred and One Dalmatians* led to a predictable increase in breeding and buying of the adorable spotted dogs, who unfortunately are also among the most challenging breeds to manage as pets.

See also Cartoon characters; Literature

Mutt

"Mutt" is a colloquial term for a mixed-breed dog. The word is often used in an affectionate or lighthearted manner. The etymology of the word is not entirely clear, but it may have originated in the late nineteenth or early twentieth century as a shortened form of "muttonhead," which was a derogatory term used to describe a foolish or stupid person. Over time, "muttonhead" might have been playfully applied to mixed-breed dogs due to their lack of pedigree, or because they were seen as less valuable or sophisticated than purebred dogs. A dictionary of English dialect has "*Mutton!*" used in scolding a dog.

See also Cur; Dog; Hound; Mixed-breeds; Purebred

N aming

A peculiar behavior of humans, in relation to dogs, is the practice of giving dogs names. There is practical value in giving each dog an individual name because it enables us to call them and train them. Naming also carries emotional valence—we name things to which we feel a special attachment. Dog naming follows broad cultural and linguistic patterns, and so it can be a window into human experience. Many of the names found in the fifteenth-century *The Names of*

All Manner of Hounds—a list of 1,065 names for hunting dogs—would sound strange to modern ears, despite their charm: Crapawde, Quarell, Flaterere, Cunnynge, Pretiboy, and Beste-of-all. So, too, might the dog names sprinkled through the plays of Shakespeare: Clowder, Merriman, Silver, Belman, and Echo. The modern dog owner might use one of the many internet dog name generators, which allow you to input breed and gender and will deliver a fitting name. Inputting "female" and "mixed-breed" results in the name Luna, which appears to be the current trending popular name for female mutts. Dog names sometimes draw on history (Brutus, Napawleon, Winston Furchill), sometimes popular culture (Buzz Lightyear, Batman), sometimes food (Oreo, Bean, Barley). And sometimes they reflect a human attempt to make a play on words, at the expense of an intelligent, sentient being (Beau Dacious).

Note of historical interest: The name Fido, which almost no one uses anymore but which nearly everyone in America associates with a dog, comes to us courtesy of Abraham Lincoln, who named his beloved yellow mutt Fido (from Latin *fidelis*, "faithful, true, trusty").

Dog names also serve as records of pedigree. A dog named Ody, short for Odysseus, is on record in the *AKC Stud Book* as Sadie's Rigorous Odysseus. Mother: Sadie, Father: Rigor Mortis; Offspring: Odysseus. See if you can decipher these names of some recent AKC champions. GCHP CH Silverhall Strike Force, GCHP Pinnacle Kentucky Bourbon, GCH CH Pequest Wasabi, GCHG CH Lagniappe's From The Mountains To The Bayou, GCHS Heywire N Deep Harbor Love That Dirty Water.

After that mouthful, we might wonder: Do dogs care what their name is? Perhaps. Some names are easier for dogs to process, and this can facilitate human-dog communication. Trainer Patricia McConnell suggests that two-syllable names might be better than one, especially in the context of recall: the first syllable ("Ohh") gets the dog's attention, the second ("Dee") is their cue to head back. McConnell says that some consonants produce a more "broadband" sound; they have more acoustic energy, more pizazz. (Sounds come out of our mouths with different frequencies and amplitudes.) Sharp-sounding consonants—those produced with force of breath, such as P, K, T—may be easiest for a dog to distinguish. Also, she warns, don't choose a name that sounds too much like a training cue (Jay/stay; Crumb/come, etc.).

Some names that humans choose for dogs are demeaning. We won't dwell here, but examples are easy to find: Fuzzybutt, Paw Pot, Shithead. Dogs don't know the difference, but it degrades them in our eyes and that's uncool.

Neoteny

Changes in the timing of development are one of the most consequential mechanisms through which evolutionary change occurs. One type of change, referred to as *neoteny*, is a delay in the rate of somatic development while sexual maturity stays on course, leading to the retention into adulthood of juvenile traits. One of the scientific debates about dogs has centered on the role of neoteny in dog behavior. For example, scientists have argued about whether behavioral traits in certain breeds of dog—such as the retention of only the first part of the predatory sequence in herding dogs, who stalk but do not kill—can be explained by neoteny. Neoteny is a form of paedomorphosis.

See also Domestication; Noses; Paedomorphism; Predatory sequence

Noise sensitivities

Along with separation-related problems, noise sensitivities and noise phobias have become a canine public health crisis among pet or homed dogs. In survey studies of dog owners in the U.S., at least half indicated that their dog had at least one behavior suggestive of noise sensitivity. Some experts believe that even this is a gross underestimation of the problem.

Noise sensitivities, as understood in the veterinary literature, encompass a range of anxious or phobic responses to sounds. Which sounds are experienced as aversive depends on the individual dog, but those that tend most often to elicit a fear response are loud and sudden, like fireworks, thunder, gunshots, and cars backfiring. Phobic reactions to a particular sound can result from a single traumatic event or repeated exposure to stimuli that a dog finds aversive. A fearful reaction to noise can manifest as panting, shaking, salivating, cringing, clinginess, all the way to a state of panic so extreme that a dog defecates on himself, chews through a wooden doorframe, or jumps out a window (all of these are true stories). A more sustained, low-level trigger is traffic noise, which is unfortunate given how many dogs live in big cities.

Dogs can also develop a sensitivity to noise in general, a heightened state of arousal even in the presence of the "normal" acoustic environment of the human home—which, let's be honest, can be a lot, with our TVs, phones, beeping appliances, vacuum cleaners, and Spotify '80s rock playlists. On top of this, many pet dogs are also subjected to the assault of ultrasonic training devices such as bark collars and fences, which emit sounds aversive to dogs (that's the whole point), but which we can't hear (that's also the point) and may grossly underestimate the impact of.

Noise sensitivities are on the rise, according to veterinary research. Why? For one thing, we live in a noisier world than ever before. There is good evidence that ambient noise levels, especially in urban areas, cause chronic stress in humans, and we have no reason to think that ambient noise wouldn't have a similar effect on dogs. Indeed, with their sensitive hearing, our noisy world may be much harder on dogs than it is on us. Dogs experiencing chronic low levels of acoustic stress may more quickly reach a threshold at which phobic reactions will occur. The epidemic of noise sensitivity may also be linked with an epidemic of untreated pain. Research has suggested that millions of pet dogs experience chronic untreated pain (from osteoarthritis, dental disease, and gastrointestinal distress, among other things). Underlying pain can lead to noise sensitivity, and noise, in turn, can aggravate pain. Dogs with noise sensitivities are also at risk of developing separation anxiety, a welfare double whammy if ever there was one.

Nomenclature

The term "dog" is often preceded by a qualifier: pet, homed, loose, stray, street, village, free-roaming (or free-ranging, which is used synonymously), and feral. Qualifiers define dogs in relation to humans, denoting dogs' level of dependence on, proximity to, and engagement in friendly relations with us.

One frequently used categorization breaks dogs into four groups: owned-restricted, owned-unrestricted, stray, and feral. These categories are useful because they remind us that some owned dogs have very little freedom to move about while others are almost completely unrestricted in their daily movements. What might not be perfect about this categorization is that it doesn't distinguish between owned and free-roaming, or between urban and rural free-roaming dogs (whose behavior and ecology are unique), nor does it make distinctions between stray, street, and village dogs—ways of life that some scholars consider quite distinct. Moreover, the term "stray" carries negative connotations—as a shortened form of "astray," it suggests that dogs in this category have strayed from home. A stray is a domesticated animal found wandering or lost (presumably away from where they belong). Yet many a stray dog would likely insist that they are not lost at all but know exactly where they are.

The lack of consensus on terminology makes it hard for scientists to compare the results from studies of behavior and cognition.

See also Feral; Free-roaming; Homed; Pet; Stray; Streeties

Noses

Dogs are a *macrosmatic* species, evolved to rely heavily on their keen sense of smell. In macrosmatic species (dogs, bears, pigs) a larger proportion of the brain is dedicated to olfactory processing than in microsmatic species, such as the human, or anosmatic species such as the narwhal. One hypothesis about the evolution of macrosmatic animals is that they evolved during a time when mammals were prey to dinosaurs. Dinosaurs were mainly visual hunters and hunted during the day. So, evolution would have favored the development of sensory capacities that would allow animals to find food and each other in low light. Smell and hearing would both have been useful under these conditions.

The sense of smell plays a fundamental role in mammalian social and sexual behavior, making the nose a valuable piece of real estate. Noses are used for sniffing and other investigative behaviors and are a key part of how dogs find food. Recent research suggests that dogs' noses can detect thermal radiation, allowing them to sense the body heat of mammalian prey, and adding to the already incredible list of things canine noses can do.

The part of the nose most recognizable to those who live with a companion dog is the nasal planum, an area of thickened, hairless skin, often pigmented black, and found at the tip of the muzzle. The ridged pattern on the epidermis of the nasal planum is unique to each individual dog, like a human fingerprint. Dogs often flare the muscles of their nostrils when taking in odors. The planum is often wet, and "wet noses" are beloved by dog owners and taken, by folk belief, to be a sign of canine

NASAL PLANUM

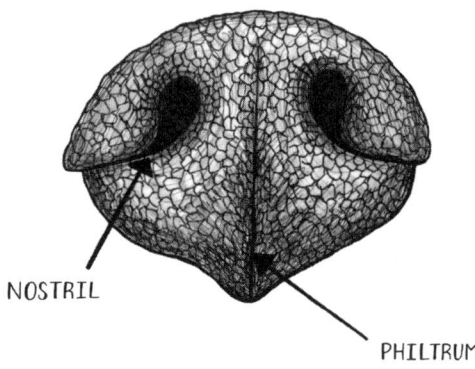

NOSTRIL

PHILTRUM

health. Although not a diagnostic symptom, a dry nose can indeed be a nonspecific sign of illness. The wet black part of the dog's face is only one piece of a larger network that makes up olfactory processing, including the brain.

Nose and muzzle are often taken to be the same thing but are distinct. The nose is the sense organ; the muzzle is the skeletal structure. The shape, length, and width of the muzzle is a key part of skull anatomy. Muzzle and skull shape vary significantly among dogs, mainly thanks to artificial selection pressures. The so-called dolichocephalic breeds (collies, greyhounds, dachshunds) are those with relatively long muzzles and heads; mesocephalic dogs are moderately proportioned (golden retrievers, beagles, chihuahuas); brachycephalic dogs are short-headed and short-nosed. In dogs with extreme brachycephaly such as pugs and bulldogs, the palate is too long for the head. When these dogs breathe, the palate flutters in the respiratory air stream, creating a noisy, labored sound. Problems associated with brachycephaly extend beyond the muzzle. The squished shape of the skull puts pressure on the brain and eyes and can cause hydrocephalus (water on the brain) and prolapsed eyeballs.

The advanced sniffing skills of dogs are extraordinarily useful for humans. Dogs are employed in over 30 different sniffer job specialties, from helping biologists find scat of critically endangered animals to helping law enforcement identify counterfeit money.

See also Communication; Conservation impact of dogs; Olfaction; Sniffing; Working

Olfaction

The canine nose is evolutionary artwork at its finest. Unlike the human nose, which is relatively small and tends to detract from our otherwise attractive faces, the nose is dogs' most prominent facial feature and the centerpiece of dogs' beauty. Indeed, no nose is too big for a canine face. Just as dogs' noses are much bigger than human noses, so too is the olfactory center of the canine brain; proportionately more of the dog's brain is dedicated to processing olfactory information than a human's brain. To put this in terms that number nerds will appreciate, dogs have 125 million to 300 million olfactory receptors compared to our measly 6 million, and the brain region where these signals are interpreted is about 40 times larger than ours. On average, then, the dog's sense of smell is about a thousand times more sensitive than ours—such an enormous difference that it is likely hard for humans to adequately imagine what olfaction truly means for dogs. Some amazing things dogs can do that we can't: dogs can track multiple smells at the same time and can adjust on the fly for wind direction and odor aging. A dog's nose sends incoming air into two separate paths, one for breathing and one for smelling. Humans have no choice but to smell and breathe through the same pathway. Dogs can detect, by smell, the presence of viruses, bacteria, and some kinds of cancer in a human's body.

Olfaction plays a significant role in a dog's experiential world. Even in utero, puppy fetuses are being exposed to chemosensory stimuli—for example, through what their mother eats—and these early experiences may shape later smell and taste preferences, just as humans begin to develop taste preferences while they are still fetuses. Puppies develop olfaction at about two weeks of age, and then begin soaking up information by sniffing everything within reach, a behavior that continues throughout life. And dogs are *always* gathering olfactory data, not just when they have their noses pinned to the ground tracking a scent, but even when they're just standing around looking like they aren't doing much of anything. Among the things dogs learn from their noses is whether other dogs have been in the area and how recently; whether these dogs are familiar or unfamiliar; whether these dogs were female or male and, if female, whether they are sexually receptive. Dogs might even gather information about the emotional state of passing dogs from olfactory cues carried on currents of air.

Dogs' olfactory work is aided by their "second nose," a structure located between the oral and nasal cavity, near the vomer bone (the hard lower end of the septum that divides the nasal cavity). Jacobson's organ, a.k.a. the vomeronasal organ, is thought to process chemical messages borne by pheromones.

One of the questions that has perplexed scientists is whether dogs have a sense of the passage of time. Does a pet dog, for example, know the difference between being left alone for 30 minutes and being left alone for three hours? Nobody really knows, and probably we never will know, but some canine researchers have suggested that dogs may "smell time" through changes in the strength of odors as they evaporate. How faint a smell has become may indicate how much time has elapsed.

See also Communication; Jacobson's organ; Lateralization; Noses; Sniffing

Owner

A common term used to describe humans in relation to other animals, especially animals kept as pets or used in industry, is "owner." In the U.S. and some other countries, dogs and other pets are legally defined as the property of a human owner, so the term is technically correct when speaking in legal terms. But many people object to the use of the word when speaking of dogs in moral terms because, the thinking goes, humans have no inherent right to make property out of other living beings. "Owner" is considered a less progressive designation than, say, "guardian." By describing ourselves as "guardian," we shift the focus onto our responsibility to provide care for and do what is in the best interests of the animal. We are invited to recognize animals as separate entities with their own rights and interests and deserving of our respect. "Guardian" has even been proposed as a legal alternative to "owner," and a few jurisdictions around the U.S. now use "guardian" language. "Owner" is inextricably tied to "pet," and so those who use "guardian" to refer to the human half of a human-dog partnership also tend to refer to dogs not as "pets" but as "companions" or "animal companions" or "companion animals."

Some industry groups, including the American Kennel Club, resist the shift to "guardian" language because, they say, it weakens the "rights" of dog owners to do what they want with and to their dog. Counterintuitively (and somewhat perversely), industry groups argue that dogs with "guardians" will have fewer protections than dogs with "owners."

See also Companion animal; Pet

Paedomorphism

Cuteness in offspring serves a key evolutionary function of eliciting a caregiving response from adults. Ethologists have described a "baby schema"—a collection of infantile features such as a round face, big eyes, a little nose, soft skin or fur, unique smells (puppy breath!), and crying sounds—that releases innate caregiving behaviors. The baby schema triggers a flood of hormones in the adult brain and, more importantly, captures attention and propels into top priority those movements that respond to the baby. In dogs and other domesticated animals, *paedomorphic* traits (from Greek *p[a]edo*, "boy, child" + *morphē*, "form") are retained into adulthood. These paedomorphic traits make dogs appear more baby-like than wolves, even as adults. The traits appeal to humans, make dogs less threatening to us, and stimulate our caregiving impulses. As you can

see by looking at a book that features photos of various dog breeds, some have more strongly paedomorphic features than others. For example, a shih tzu is "baby-faced" while a German shepherd has a more "wolf-like" face.

See also Neoteny

Paleofeces

Paleofeces, or fossilized dog poop, is an important source of archaeological data. Ancient poop can tell us a lot about when, where, and how a species of animal lived. Like a crowded hiking trailhead, the archaeological record is dotted with generous amounts of dog poop. Among other things, researchers are hoping that ancient dog poop will provide new insights into dog domestication, especially how the dog microbiome has changed over time, as a shift occurred from the carnivorous diet of wolves to the omnivorous diet of dogs, who shared food with humans. Distinguishing ancient dog poop from ancient human poop is surprisingly difficult: human poop sometimes contains dog DNA because people ate dogs. Dog poop, for its part, contains human DNA because dogs ate human poop. A fossilized poop is also known as a *coprolite*.

See also Defecation; Dogor; Domestication

Panting

Compared with humans, dogs have relatively few options for personal climate control. Dogs mainly regulate body temperature through panting. During panting, the moisture on a dog's tongue evaporates, cooling the body. Panting facilitates the evaporation of moisture from the lining of the lungs, which also helps bring body temperature down. In addition, dogs release heat through their paw pads, nose, and ears. Dogs have very few sweat glands, compared to humans, and most of these are in the foot pads.

Another way that dogs thermoregulate is through their pelage, or coat. Seasonal molt of the pelage—known affectionately to human dog owners as shedding—is an adaptation for coping with temperature extremes. Dogs will carry lighter-weight coats in warmer parts of the year, and then thicken up for the cooler seasons.

See also Ears; Paws

Parasites

For those of us who share intimate spaces with our canine companions, being reminded of the extensive and rather gruesome list of parasites that live in or on our dogs might be unsettling. If you dare, read on. Otherwise, maybe just skip this section.

Parasites are, as the name suggests, pathogenic organisms who live in or on a host, often at the host's expense and without compensation. The Companion Animal Parasite Council lists well over 100 parasites who may, if we share our home with a dog, also share our home. Many of these are external parasites, hunkered down in the furry coats of dogs. Ticks dig into their host animal's skin using two sets of hooks and attach their mouths so they can feed on blood for 3 to 10 days. During this intimate exchange with dogs, ticks may pass along any number of nasty tick-borne diseases, including Lyme disease (yes, dogs get Lyme), Rocky Mountain spotted fever, and canine ehrlichiosis. Mites are microscopic arachnids. Some species of mite live on a dog's hair follicles and others burrow down into the dog's skin. In both cases, mites can cause an extremely itchy and debilitating skin condition called mange. The so-called chewing lice, or Mallophaga, have large mandibles with which they grasp their canine host's hair. Sucking lice, or Anoplura, by contrast, have narrow heads adapted to sucking blood and fluids, and use their claws to hold onto hair. One species of chewing louse, *Trichodectes canis*, is a vector for dog hookworm.

Which brings us to internal parasites. Dog hookworms (*Ancylostoma caninum*) are the most prevalent nematode parasite in dogs. Once inside their host, dog hookworms use their hook-like mouth to attach to a dog's intestines, where they will enjoy a smorgasbord of blood and tissue fluids. The dog host will continue to lose weight, have bloody diarrhea, and grow weaker and weaker. When untreated, puppies can die from hookworm. Hookworm can spread quickly when dogs are kept captive in large numbers and in close quarters, as in kennels of greyhounds used in dog racing. Alarmingly, dog hookworms appear to be developing antibiotic resistance. Other internal canine parasites include roundworms, heartworms, tapeworms, whipworms, and *Giardia*, a tiny parasite that causes the unpleasant but relatively familiar diarrheal disease giardiasis.

Although not as overtly creepy as parasites, viruses and bacteria also contribute to the global canine disease burden, and influence human-dog relationships in various ways. Among the viruses of most concern for dog health is canine distemper, also called hardpad disease because it can cause hardening of the nose and paw pads. Distemper is highly transmissible through airborne particles or contact with infected surfaces such as food bowls. Canine parvovirus is also highly contagious, spread mainly through contact with the feces of a sick dog. Both illnesses can and do kill dogs. Kennel cough, often caused by the bacterium *Bordetella bronchiseptica*, is an infection in the trachea and bronchial tubes and causes coughing. The colloquial name "kennel cough" nicely captures the highly transmissible nature of *Bordetella*, which can sweep like wildfire through a population of group-housed dogs.

See also Flea control; Licking; Quarantine; Rabies

Paws

The dog pawprint is an iconic symbol of the human-dog bond, a shorthand either for "Dogs love us" and/or "We love dogs." So ubiquitous and well understood is the symbol of a pawprint that it is enshrined in an emoticon. Of course, for dogs, paw pads are more than just feel-good symbols.

The canine foot has four toes, each with three phalanges (segments). Each foot also has four digital pads, one cushioning each of the toes, and one communal pad, the

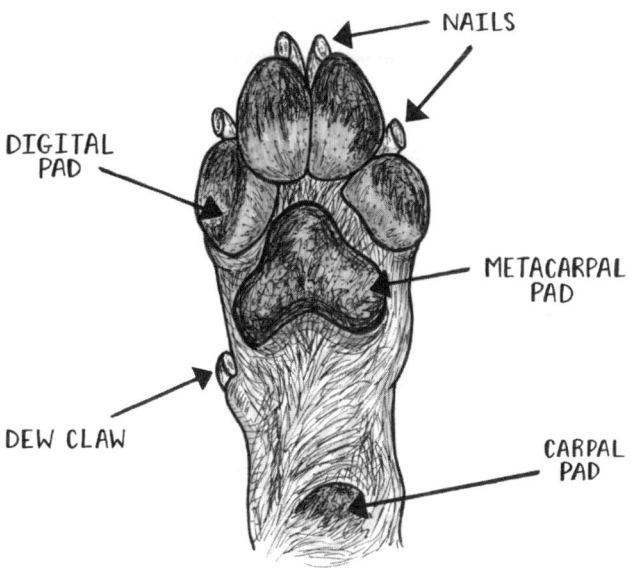

NAILS

DIGITAL PAD

METACARPAL PAD

DEW CLAW

CARPAL PAD

large metacarpal pad. Dogs are *digitigrade*, which means that their toes carry their body weight. (Humans are plantigrade: like bears, we walk on the soles of our feet.) The paw pads, then, are what cushion dogs' bones and joints during locomotion.

The paw pads are formed of adipose tissue and elastic fibers covered in a thick hairless skin, usually pigmented black. In addition to absorbing force, the paw pads function as "shoes" of a sort, protecting dogs' feet from rough surfaces, rocks, heat, and cold. Paw pads may also serve as scent spreaders, although whether and how exactly dogs use "pedal scent"—leaving a scent trail with the feet—for communication is poorly understood.

Eccrine, or sweat, glands in their paws help dogs regulate their body temperature. Dog paws are often described as smelling like Fritos corn chips. This odor comes from the eccrine glands.

Paws and pawing also function in dog behavior. Fore-paw raising can be a greeting, or a solicitation for attention or food or play. Pawing is used to communicate both with other dogs and with humans. For a dog interacting with a human, pawing can be a way to intensify begging. If a laser stare and drool dripping onto the floor are not enough, a gentle paw on the knee is a way of saying, "Come on, pleeeease share your food!"

Paws bring us to feet and foot anatomy and shape. Selective breeding has influenced the shape of dogs' feet. The "normal" canine foot is slightly oval. But some breeds have feet that are more rounded, like a cat's, or more flattened, especially wide, or especially narrow, and so forth. The extent of webbing between toes also varies by breed, with "water dogs" (e.g., Chesapeake Bay retriever, Newfoundland) having more pronounced webbing. Paw pad shape mirrors the shape of the foot, so different breeds will have distinctive pawprints.

See also Communication; Cursorial; Olfaction; Scent marking

Pedigree

Pedigree is not a biological category, but rather a human cultural label. Pedigree refers to the lineage of a particular dog who belongs to a recognized breed and whose ancestry has been (as far as humans know) tightly controlled and recorded in stud books.

See also Breed; Purebred

Pet

Human-dog relationships take many forms, one of which is companionship. And this companionship, in turn, takes many forms, one of which is a dog living as a pet. "Pet" is a loose designation, taken broadly to mean a dog who lives in a companionable relationship with a particular person and who shares a dwelling with this person. Yet the life of one pet dog may be vastly different from the life of another pet dog: some pet dogs have almost complete freedom of movement and would fit the definition of free-roaming dog. Other pet dogs are completely confined to the walls of a house or apartment, or even a small space within, and could be described as captive. "Pet" is often paired with the term "owner."

"Pet" has typically been used, even in scholarly writing, to refer to a dog who serves no utilitarian function. So, "pets" are differentiated from working dogs such as herders, sentries, and sled pullers. But this is misleading, because being a pet is also a form of canine labor, one that exploits the affiliative skills of dogs and involves the work of providing intensive emotional companionship. Indeed, working as a pet can be one of the most taxing jobs dogs are asked to do. Moreover, dogs who work in herding, hunting, guarding, and other canine professions are not valued only for the work they do; they often live in close companionable relationships with their human partners, having familial ties that are stronger in many cases than those between pet dogs and their guardians.

In relation to dogs, "pet" also functions as a verb. To pet is to stroke gently. Humans love to pet dogs, and dogs often enjoy being petted, although dogs seem to prefer some kinds of petting to others. Based on the few studies of how dogs respond to petting, it appears that humans generally want longer and more intense tactile interactions than dogs do. Research has shown that most dogs dislike being touched by humans on their paws, hind legs, and the top of their head—unfortunately, some of the spots toward which people gravitate. Most dogs would rather be petted on the side of the chest and under the chin. But individual dog preferences vary widely, and even a particular dog may have certain times he or she likes to be petted and other times when petting is unwelcome. Inappropriate, ill-timed, and invasive petting by humans is a major factor in dog bite incidents. So greater awareness by humans of dogs' preferences would make everyone happier.

See also Free-roaming; Homed; Owner; Working

Pet effect

The term "pet effect" refers generally and somewhat ambiguously to the purported physical and mental health benefits of animal companionship. According to the pet effect, the presence of a dog acts as a buffer against anxiety, depression, and other negative feelings, improves cardiovascular health, leads to improved exercise habits and to longer life. Despite the daily repetition of this pet effect claim in popular media, the nature of the association between health and pet ownership is complex, diverse, and controversial. Evidence is not straightforward or fully convincing.

The connections between pet ownership and *decreased* individual mental or physical well-being have received less attention, shadowed by the bright glare of the pet effect. Research by psychologists has suggested that pet ownership can increase levels of stress and negatively impact quality of life, especially when a person is caring for an animal who is behaviorally challenging or who is suffering from illness or disability. Pet ownership is also associated with some increased risk of physical injury, most obviously from bites and less obviously from orthopedic injuries caused by rambunctious dogs tearing around the house doing zoomies. Moreover, the pet effect, in which pets are presented as a boon to society's overall well-being, is rarely balanced against the social costs borne by veterinarians, who have the dubious honor of being the profession with the highest suicide rate.

The pet effect meme also glosses over important questions about whether and how dogs and other pets benefit from their association with people and whether and how animals kept as pets are harmed. Recent research suggests that the mental health of pet dogs is very poor; dogs suffer high rates of anxiety and perhaps also depression (though no one has yet focused attention on canine depression rates). As humans get happier from their close association with dogs, dogs seem to get less happy. The possibility of this kind of inverse relationship has not yet been rigorously studied.

See also Pet; Working

Play

Humans may not readily understand the importance of canine play, one of many "cultural" misunderstandings between human and dog. We associate play with leisure, hobbies, games. It's what you do with children. But for dogs, play is an essential component of behavior

and well-being. Play is especially important for young dogs: it is good physical and mental exercise, and it is an important part of how puppies learn to communicate with and behave appropriately around others. It helps develop their furry little bodies and their growing brains. And, of course, play is fun.

Adult dogs play, too. Indeed, the level of playfulness in adult dogs is unusual. In most species, play is mainly the province of the young. Exactly why adult dogs are so playful is subject to some disagreement. Some scholars have argued that enhanced playfulness is a byproduct of paedomorphosis. Others have argued that adult playfulness is an adaptive trait that facilitates the formation of human-dog bonds. Either way, it is something that many dogs enjoy, though some seem to enjoy it and want to do it far more than others.

Canine play is often divided into categories of social, object, and solitary play. These, of course, can be overlapping, as when one dog playing alone with a stuffy decides to initiate a stuffy tug-of-war with another dog or a human. All these categories of play involve motor patterns performed in other contexts, such as predation, agonistic encounters, and courtship. Within the context of play, these behaviors may be differently sequenced or may vary in intensity.

If you watch dogs at play and pay attention to their posture, ears, eyes, and tails, you'll quickly observe that play involves an ongoing flow of information, with the clear message: This is play, not something else. If I bite you, it is just in play. Communication is essential, because the behaviors used in the context of play, such as vigorous biting, body-slamming, and growling, could easily be misinterpreted as aggressive or agonistic, which could lead to fights and to serious injury.

As we know from the work of ethologist Marc Bekoff, dogs perform certain highly stereotyped action patterns in the context of play, and these serve as not-to-be-misunderstood signals of intention. Among the most fun to watch is the play bow, in which dogs crouch on their forelimbs while sticking their butt in the air. The play bow says to another dog that the behaviors that are on their way—a bite or a body slam or a mount—are just for fun. Other body postures and movements that signal play include "gamboling"—a loose jumping and capering—pawing, and a facial expression in which the jaws are stretched wide open, lips pulled slightly back, tongue in view. Dogs also use vocalizations to signal their playful intentions, including mock-ferocious play growls and the "play-pant," a forceful huffing of air that has been likened to laughter.

Play behavior has been hypothesized as a demonstration of moral behavior in dogs. Dogs follow a strict code of conduct when they play; they play fairly and don't cheat. They take turns being on top and will adjust their behavior to keep the playing field even by performing what Bekoff refers to as self-handicapping (both individuals will make themselves vulnerable, for example by rolling over on their back) and role reversal (larger or stronger individuals will go easy on smaller or weaker individuals). Play requires a level of trust. Rolling over on your back is a vulnerable posture for a dog; you must trust that your play partner won't attack. Play requires honest communication. Dogs who cheat—who say they want to play and then bite too hard—will be excluded from play.

See also Communication; Emotions; Growling; Inequity aversion; Joy; Zoomies

Poop laws

Doodoo, or dog poop, serves certain behavioral functions among dogs, including olfactory and visual signaling. It is also significant as a site of human-dog conflict. Poop has driven attitudes toward and regulation of dogs in public spaces, especially in urban areas. Along with bites and rabies, dog poop is the main reason we now have leash laws and why it is illegal for dogs to roam loose around cities. Many municipalities have laws governing the ownership of and responsibility for dog poop. If you own the dog, you also own the poop. Owners who fail to clean up after their dog can be fined. Poop is also one of the most important motivators behind "dog rage" incidents: dog poop has led to several fatal shootings in the U.S., either of the offending dog or of the offending dog's owner.

Dog poop is a legitimate problem, with environmental, social, and public health ramifications. These problems are particularly acute in large cities, and especially in countries with high numbers of pet dogs. Because of the way we feed pet dogs (lots of animal protein), dog excrement has high levels of nitrogen and phosphorus. Dog poop also contains high levels of fecal coliform bacteria and sometimes transmissible parasites. And because there are so many dogs, there is an enormous quantity of excrement. It can make people and other dogs sick and can disrupt ecosystems and put pressure on native plant species.

See also Cultural attitudes; Fake turf; Shelters

Predatory sequence

Even though many dogs don't have to hunt because their food is supplied by humans, they still have predatory behaviors, defined as behaviors by which an animal of one species finds, catches, kills, and eats an animal of another species. A topic of considerable interest among evolutionary biologists who study dogs is whether and to what degree dogs retain predatory motor pattern sequences found in their wild ancestors, particularly the wolf.

The basic wolf predatory sequence goes like this:
Eye
Orient
Stalk
Chase
Grab/bite
Kill/bite
Dissect
Consume

Do dogs engage in all these behaviors, in this order? No one seems to agree. Heritable behaviors derived from the wolf predatory sequence are certainly retained in modern dogs. To offer an obvious example, when we throw a frisbee for our dog, we are hoping to trigger predatory behavior, particularly the chase drive. But dog predatory behaviors have been under direct selection pressure for thousands of years, and therefore show variation in the contexts within which behaviors take place, the sequencing of behaviors, and the thresholds for behavioral response.

Research has suggested that the predatory sequence varies among dog breeds, depending on their work specialization. Some dogs have been bred for an intact predatory sequence, while in others, certain components of the sequence have been exaggerated. The chase drive is particularly strong in sight hounds, while the kill/bite component is strong in terriers, who have been bred to catch and kill rats. In other cases, some components of the sequence may have been inhibited. For example, in dogs bred to guard livestock, the kill/bite component has been inhibited, since the idea is for dogs to protect, not try to eat, their flock. The characteristic pointing stance of the so-called pointers, like the German shorthaired pointer, is the exaggeration through selective breeding of a natural stance. All dogs point. Predators approach prey slowly, without making a sound, eyes locked on prey, each leg lifted (point!) ever so slowly to reduce the appearance of movement.

See also Domestication; Head shaking; Neoteny

Puppy breath

The sweet-smelling breath of puppies sends humans into paroxysms of delight. The sweet smell is likely a combined effect of mother's milk (although this doesn't fully explain it, because the sweet breath continues past weaning), the bacteria present in the puppy's gastrointestinal tract, and clean teeth (the absence of bad breath from poor oral health). The distinctive smell generally fades by six months of age, as puppies get their adult teeth.

See also Teeth

Puppy-dog eyes

One of dogs' most striking paedomorphic features is colloquially known as "puppy-dog eyes," a distinctive facial expression seen, especially, on dogs who are trying to separate us from the food we're about to put into our mouth. Creating puppy-dog eyes involves the use of a specialized muscle called the *levator anguli oculi medialis*. Scientists have compared the facial musculature of dogs with that of wolves and have found that the LAOM muscle is far more pronounced in dogs than in wolves. It is strong enough to pull the eyelids up, exposing the

sclera, or white of a dog's eye, at the lower eyelid. The eyes appear larger and more paedomorphic, and a bit sad, thus hopefully triggering a caregiving response from us (giving food to the poor, starving baby). Puppy-dog eyes should be distinguished from so-called whale eye (also called half-moon eye), which also involves the eye whites. A dog looking sideways, eyes wide and whites visible, is showing signs of worry.

See also Begging; Domestication; Facial expressions; Solicitation behavior

Purebred

"Purebred" is a rather distasteful concept, with its grating eugenic overtones. And in truth, purebred doesn't mean much from a biological point of view. Genetics is so messy and complicated that "pure" doesn't make a lot of sense. Purebred is a social construct used to describe and define dogs and, it must be said, to fan the flames of the overheated and highly profitable market for pet dogs. With those qualifications, here is the official definition: a purebred dog is an individual belonging to one of the 300+ "official" dog breeds in the world, as recognized by international dog breeding organizations such as Fédération Cynologique Internationale or the American Kennel Club. To be purebred, a dog must have documentation proving that both parents were of the same breed and must adhere to highly specific morphological qualifications. For example, the dachshund must have a ground clearance one-third of the height at withers, and the proportion of the body length (measured from the manubrium, or uppermost segment, of the sternum to the point of the buttock) to the height at withers should be

1.7:1. Ears should reach the edge of the lips, eyes should be almond shaped and dark reddish to blackish brown, and tail should be "carried in a harmonious continuation of the topline" (as per FCI breed standards).

See also Artificial selection; Breed; Mixed-breeds

Qartuli nagazi

Yes, we are stretching for a "Q" word. But the gorgeous Georgian or Caucasian shepherd (transliterated from Georgian ქართული ნაგაზი as *qartuli nagazi*) is nevertheless worth mentioning. One of the largest breeds of dog, these shepherds weigh between 80 and 180 pounds, making them well suited to protect livestock from wolves and bears. Researchers have been studying the mitochondrial DNA of Georgian shepherd dogs to explore the hypothesis that local Caucasian wolves have contributed substantially to the dogs' gene pool, through human-mediated hybridization (pairing a dog with a captured wolf) and through naturally occurring hybridization.

See also Breed; Hybrids; Landrace

Quarantine

The word "quarantine" dates to the fourteenth century and refers to the Venetian policy of keeping ships from plague-stricken areas at port for a period of extended isolation—typically 40 days (*quaranta giorni*, "space of forty days")—before allowing them to unload, ensuring that no latent cases of the bubonic plague were on board. Like ships at sea, dogs are vessels that can carry disease; when traveling overseas, dogs are sometimes forced by humans to undergo periods of isolation, or quarantine. Many countries require a dog traveling across their border to quarantine at a designated facility, to confirm that the dog is free of infectious diseases, especially, but not only, rabies. In the U.S., quarantine is not required for pet dogs entering the country; however, the Centers for Disease Control requires health certifications proving that dogs don't have rabies. Regulations for entry into the country are particularly stringent for dogs who have been in a country with high risk for rabies within the past six months. Canine rabies was eliminated in the U.S. in 2007, but since 2015 several rabid dogs have been imported into the country, raising concern about a resurgence of the disease.

In the U.S., all dog bites must be reported to state authorities. Dogs who have bitten a person or another dog must be quarantined by their owner or in an official quarantine facility for ten days after the bite, a so-called bite quarantine. If a dog has been bitten by another dog whose rabies status is unknown, the bitten dog must be quarantined, even if vaccinated. Quarantining a dog involves an owner keeping the dog on his or her property or on a leash if off the property, usually for ten days. Sometimes quarantine will be recommended by a veterinarian for other reasons, for example if a dog has parvovirus, canine influenza, Bordetella, or giardia—diseases that are highly transmissible.

See also Parasites; Rabies

Queen Elizabeth pocket beagle

The search for a "Q" word also provides an opportunity to highlight the morally questionable practice of inventing dog breeds that have no functionality other than to serve human whims and fancies. Because regular beagles were not small enough for everyone's taste, some enterprising breeders decided a few years ago to market a beagle that is extra-small and can fit into one's pocket. The Queen Elizabeth pocket beagle is half the height and weight of a standard beagle, weighing in at just 5–8 pounds. At 5 to 11 inches in height, the dogs are small enough to ride in a purse. The QEPB is part of a contemporary trend toward the miniaturization of pet dogs. Consumer demand for small dogs is at an all-time high, and small-minded breeders are busy creating "pocket" and "teacup" versions of many already-small breeds of dog, sometimes by mating runts and sometimes by stunting the growth of puppies to keep them small.

See also Selective breeding; Small Dog Syndrome; Yappy dogs

Rabies

Rabies has shaped human-dog relationships throughout our shared history, and the disease continues to be one of the most significant forces in the complex biological, social, and cultural interactions of humans with dogs. Although rabies has been virtually eradicated among dogs in some places, including the U.S. and the U.K., it imposes a huge disease burden around the globe, especially in countries where rabies vaccination rates are low and where high numbers of feral and free-roaming dogs live alongside humans. The World Health Organization estimates that between 30,000 and 70,000 people die of rabies each year, with almost all these deaths occurring in Asia and Africa. How many dogs die of rabies? It is hard to know. If you ask Google, "How many dogs die each year from rabies?" you will be routed to reports on human mortality. But the disease burden for dogs is likely significant.

Rabies is a *zoonotic* disease (a disease transmitted between animals and people) caused by *Rabies lyssavirus*, a single-strand RNA virus. It is spread through saliva, particularly after a bite, as infected saliva enters a wound. It can affect all mammals, but the most common route of transmission to humans is through dogs, simply because people are in much closer contact with dogs than with skunks, bats, and racoons. Indeed, 99 percent of human rabies infections are caused by dogs, and particularly by dog bites. It is worth cautioning that feral and free-ranging dogs are no more likely to bite than pet dogs. Indeed, although comparisons are fraught and data are slim, bites are probably far more frequent in the domestic setting, where people and dogs are in extreme proximity and where dogs are often placed in stressful situations. Still, because pet dogs are almost all vaccinated, the disease risk from bites is relatively low.

Written accounts of dogs infecting humans with rabies date back roughly 4,000 years, making rabies one of the oldest known zoonotic diseases. A Mesopotamian code of conduct called the Laws of Eshnunna (ca. 1930 BCE) warns citizens that if the owner of a rabid dog doesn't watch over his dog properly and someone gets bitten and dies, "the owner of the dog shall pay forty shekels of silver." If the dog bites a slave and causes his death the owner "shall pay fifteen shekels of silver."

Rabies was and still sometimes is referred to as hydrophobia, as in the popular 1957 movie *Old Yeller*, in which a boy's beloved dog (a yellow cur) is bitten by a rabid wolf. Yeller develops the "hydrophobie," requiring the devastated youth to shoot his own boon companion. The moniker relates to one of the most bizarre symptoms of rabies infection—and there are many: an intense fear of water.

Countries that have achieved low rates of rabies transmission have done so through aggressive vaccination and control campaigns. In the U.S., for example, you cannot legally take your dog out of your house unless he or she has a rabies vaccine certification and tag. Dogs who have bitten someone are quarantined, and dogs coming in from other countries must have proof of vaccination status. Although these austere tactics have been successful in combating rabies infections, dogs have paid a high price. The late nineteenth and early twentieth centuries were a particularly difficult time for dogs in the U.S., U.K., and some European countries, as loose and unowned dogs were rounded up by the hundreds of thousands and killed, often in violent and inhumane ways.

See also Cynoctone; Movie stars; Parasites; Quarantine; Shelters; Stray; Streeties

Racism

Within the modern Western (white, upper- and middle-class) cultural imagination, humans and dogs—or more often Man and Dog—are depicted in a loyal, mutually beneficial, affectionate relationship. People who don't like dogs or who are afraid of dogs are viewed with some suspicion. How could you not love dogs? Yet there are good reasons why individuals and even certain groups of people might have complicated and not entirely positive feelings toward dogs. Rabies gives one view into this: for many people, now and in the past, dogs are associated with a dreadful disease and are a source of fear, not succor. Another view is from the complicated entanglements of dogs and racism.

One of the most overt and brutal links between race and dogs has been the weaponizing of dogs as a tool of racial oppression. To offer just a few of many possible examples, dogs were used to enforce captivity on Blacks in the antebellum South and were used as weapons during the appropriation of Native American lands by colonialists. Dogs have also been tools of anti-Semitism and were employed by the Nazis to terrorize Jews held in concentration camps. Dogs—especially the use of dogs in law enforcement—are still racially charged in contemporary America. During the 2016 demonstrations in North Dakota against the Dakota Access pipeline project, which cuts through large swaths of the Standing Rock Sioux tribe's sacred land, security officers contracted by the construction company brought in dogs to intimidate and attack protesters.

Charles Darwin wrote a lot about dogs, and one of the things he pondered (in *The Descent of Man*) was the possible similarity between human races and dog breeds. Science and politics since have moved us to a place of greater understanding, and Darwin's question now reads as deeply uncomfortable. Yet an analogy between race and breed has long exerted a kind of magnetic power and has been damaging to people and dogs alike. Scientific theories of racial purity promulgated from the mid-nineteenth through much of the twentieth century were applied to both dogs and humans. It is perhaps no coincidence that dog pedigreeism—the idea that certain breeds are superior to others and, especially, that being of "pure" breed infers value and status and being a mutt or mongrel makes you worthless—was gaining popularity in the U.S. at the same time as negative attitudes toward immigrants (Irish, Italian, Jewish, Chinese, etc.) reached their apogee. Early-twentieth-century laws against miscegenation and the forced sterilization of genetically "unfit" people also reflect persistent obsession with notions of racial purity. The opposition to "random" (not selected and prescribed by humans) mating by dogs is still with us, and stereotypes about dogs of mixed or uncertain origin—especially that they are more likely to be behaviorally troubled and "bad" than purebred dogs—remain affixed like barnacles to the hull of the human imagination.

One other facet of the complex discussion of race and dogs is the linking of racism and speciesism. In contemporary political discussions, the treatment of animals—especially in agriculture and research, but also in dog-breeding facilities and even pet-keeping—has been likened to slavery, and "speciesism" (the discrimination against certain kinds of beings based on their species membership) likened to racism. Although most scholars find the analogy problematic, some have nevertheless used it to develop conversations about "intersectionality," which is the idea that different forms of oppression—racism, speciesism, and sexism—are linked and entangled. In addressing them as linked concerns, we may be able to better understand and untangle these forms of oppression.

See also Weaponizing; Xenophobia

Red Rocket

Humans are uncomfortable around dog penises, which is why this topic is listed under the sneaky heading of "Red Rocket." Nearly everyone with a male dog will either know the colloquialism "Red Rocket" or will have some other nickname for It. Lipstick. Pink Thing. Rod of the Rovers. Canine Incher. All refer to the red menace lurking beneath all that innocent-looking hair.

The furry, oblong sheath dangling in the nether region of the male dog's belly is not actually the penis itself; it is the prepuce. The prepuce, or penile sheath, plays the vital role of protecting the penis from assaults of weather and other potential trauma to which the Little Guy might be exposed. The penis itself lies hidden—mostly!—within the prepuce. But it makes an appearance here and there.

When and why does the Red Rocket show itself? Physiological arousal. The arousal is often sexual in nature, but any kind of excitement can lead to a partial erection and, often, to humping. Anatomical features unique to each dog determine the frequency and obviousness of Red Rocket appearances. For example, some dog penises do not fit perfectly into the prepuce and the tip of the Lipstick pokes out. Even more disturbing to some humans, though completely natural and helpful to the dog himself, is the yellowish fluid called *smegma* that coats the penis and may accumulate at the tip of the prepuce. Smegma keeps the penis lubricated and healthy.

Neutering may make Red Rocket appearances slightly less frequent, but because arousal isn't always limited to sexual encounters, the Red Rocket will still come out.

The canine penis has a bone in it; in the middle of the bone is a gland called the *bulbus glandis* that swells during sex. This swelling functions to keep the penis inserted in a female dog's vagina, helping ensure that ejaculation is complete before withdrawal. At the same time, a muscular band in the female dog's vagina constricts, further holding the penis tightly in place. This so-called "copulatory tie," during which female and male dog are stuck together in the lovemaking dance, can last up to 30 minutes.

Male dogs can suffer from a medical condition called *paraphimosis*, or pathological extrusion of the penis from the prepuce. (Uncircumcised male humans can also suffer from paraphimosis if the foreskin becomes trapped behind the glans.) In simple terms, the Red Rocket has come out and has gotten stuck and can't get back inside again. Paraphimosis has a variety of causes, one of the most common of which is that the hairs at the tip of the prepuce have clogged the opening, preventing the penis from retracting. Paraphimosis is often precipitated by manual semen collection by humans for breeding purposes or can occur after misdirected sexual activity, such as humping a pillow or a human leg. The condition can cause swelling and drying of the penis and can be painful. If extrusion continues for an extended period, it can result in serious tissue damage. Ouch. Call the vet! (A veterinary diagnosis of paraphimosis would be made if the penis is extended from the prepuce for more than 2 hours.)

See also Mating; Reproduction

Reproduction

Dog reproduction has been altered in interesting ways by the process of domestication. Domestication pressures on dogs have compressed the time frame for sexual maturity. Female domestic dogs reach sexual maturity during their first year of life, typically at around 9 months of age. This is earlier than female wolves, who don't reach sexual maturity until about 22 months of age. Nearly all female dogs go through two cycles of receptivity a year, whereas wolves typically go through one. Likewise, male dogs reach sexual maturity as early as 5 months old, although their fertility generally doesn't peak until after one year of age. Male wolves don't reach sexual maturity until they are at least two years old. Male dogs are promiscuous, which means they are willing and eager to mate with any sexually available female. Wolves, in contrast, are monogamous, and often mate with the same partner for life. Taken together, these changes to reproduction have allowed humans to breed more dogs at a faster pace, speeding up selection for certain desired traits.

See also Mating; Red Rocket; Sterilization

Rolling

Rolling serves various functions, most of them related to scent. A dog rolling on the ground is either trying to absorb odors or to leave odors. The behavior is also sometimes called scent-rubbing. A dog releases odor into the ground by rubbing face and body—performing a kind of full-body marking. Rolling may also be directed at absorbing odors. A dog might sniff a spot on the ground, maybe even give it a little scratch to release odors, and then drag head and shoulders and then back and butt through the special spot. Rolling may be an attempt to camouflage their own smell, to blend in with the environment, smell-wise, and be less noticeable to prey. Rolling in stuff may also leave a dog smelling good (in their own opinion). Just as we might wear jewelry, hats, or coats—and apply perfume or shaving lotion—to express our individual creativity and tell others a bit about who we are, so too do dogs adorn themselves with odors.

As many dog guardians will attest, dogs will often roll in the vilest substance available (dead fish, feces, rotting deer carcass), perhaps because it smells appealing to them, or maybe just for the sick pleasure of watching our hysterical reaction. Sometimes, dogs roll simply because it feels good, like a satisfying back scratch.

See also Olfaction; Scent marking

Sagittal crest

The sagittal crest is a ridge of bone that sits on the top of the skull, at the junction of the two parietal bones. The sagittal crest is present in dogs and other canids, along with many other vertebrate species who have evolved to chew tough foods; the presence of a sagittal crest indicates strong jaw muscles. (Humans don't have a sagittal crest.) Scientists studying the size and shape of dog skulls are interested in how domestication may have changed the shape of the head, along with dogs' brains, teeth, eyes, and noses. In trying to identify fossil remains, the shape of the skull is a key factor that helps scientists decide whether they are looking at a dog or a wolf or some in-between ancestor. Among the features of

SAGITTAL CREST

skull morphology that differentiate wolves and dogs is the size of the sagittal crest. The larger sagittal crest of wolves allows for a much more powerful bite. Some scientists speculate that domestication may have led to a reduction in the size of the sagittal crest in dogs, along with a reduction in the power of their bite, because dogs were trained to bring prey back to humans rather than consume what they caught. All dogs have a sagittal crest, but the bump is especially prominent in some breeds, typically those with long noses (e.g., vizslas, collies, Dobermans).

See also Noses; Wolves

Saliva

A.k.a. slobber. Drool. Saliva plays many important roles in the life of a dog: it helps maintain homeostasis in the canine oral cavity, and it lubricates the oral mucosa, which is important for maintaining healthy teeth.

Saliva aids dog digestion by moistening food and helping with bolus formation, smoothing the way for food to make its way down the esophagus. (A bolus is a soft mass of masticated food.) As you may have noticed if you live with a dog, dogs often don't chew their food, they vacuum it up. Human children are instructed to chew each bite of food 32 times—and for humans, chewing is an important part of digestion. Saliva mixes with food in the mouth and contains enzymes that begin the process of digestion by breaking down carbohydrates and fats. Dogs don't begin digestion until the food is in the stomach and intestines, so wolfing down their food isn't really a problem.

Dog saliva helps prevent doggie cavities—more effectively, we might note, than human saliva prevents human cavities. Dog saliva—and the saliva of carnivores in general—has a pH of 7.5–8, which is slightly alkaline. Human saliva has a pH of 6.5–7. The alkalinity of the dog's saliva serves to buffer the acids produced by bacteria in the mouth. Acids can wear away at tooth enamel, making teeth more susceptible to cavities.

Contrary to a popular myth propagated by dog lovers, a dog's mouth is not "cleaner" than a human's mouth. A mouth is considered "dirty" if it contains bacteria that might cause disease to someone else. Dog and human mouths are equally disgusting: both species have over 600 different types of bacteria that can live in their mouth. (Only about 16 percent of these microbes are common to both species.) Dog saliva can transmit disease to humans, one pathway for zoonosis.

As in humans, excessive saliva production can be a sign of ill health in dogs. You've probably had the uncomfortable experience of having your mouth fill up with saliva just before you vomit. Hypersalivation can be caused by nausea, as well as by a host of serious medical problems. What's "excessive" is hard to define; some dogs are naturally more slobbery than others.

Dogs and salivation are indelibly linked with Russian researcher Ivan Pavlov, who studied the salivary system using dogs as a model. To understand how eating was related to gastric, salivary, and pancreatic secretions, he developed a model of sham feeding: he would surgically remove dogs' esophagi and cut a hole in their throat, creating what is called a gastric fistula. When the dogs ate, the food would fall out of the hole. Pavlov attached a little container to the fistula to collect the gastric and salivary fluids produced during feeding. Gastric fistula surgeries are still performed on dogs in medical research laboratories today.

See also Brown Dog Affair; Flews; Parasites; Teeth

Scent marking

Dogs do a great deal of communicating through scent and olfaction. Like wolves, dogs use olfactory signals—including pee, poop, and pheromones from scent glands in the paws and near the anus—to send messages to other dogs about who they are, where they have been, and how they are feeling. Using urine, dogs leave information about identity, rank, and reproductive status. Dogs also scent mark with feces. Dogs will sometimes scratch the ground after pooping or peeing, to add a visual mark and accentuate their message. Scent marking shapes both agonistic and affiliative interactions between and among dogs. Dogs use scent to defend resources or territory, advertise availability, or leave personal information for other dogs. (Honestly, there is a lot going on with scent marking that human scientists simply don't understand.) The ability to scent mark is constrained for many pet dogs by homes and leashes and fences, and they are expected to pee and poop in a narrow range of places chosen by their human.

See also Communication; Defecation; Olfaction; Rolling; Sniffing; Territoriality; Urine

Scott and Fuller

One of the keystone books about dog behavior, and the basis for a great deal of contemporary puppy-raising advice, is one you've probably never heard of: *Genetics and Social Behavior of the Dog*, by John Paul Scott and

John H. Fuller. Scott and Fuller conducted long-term and detailed studies designed to increase our understanding of the genetic basis of dog behavior, and to tease apart the influences of genetics and environment on the behavior of dogs. They tested five breeds, representing major breed groupings, for traits such as aggressiveness, trainability, problem solving, and emotional reactivity, among other things. (In case you are interested, the five breeds were beagle, sheltie, fox terrier, basenji, and cocker spaniel.) They also did systematic crossbreeding to see what behavioral traits would be genetically passed down. What they found, in short, was that "nurture" (environment) had at least as much importance as "nature" (genes, heredity), if not more.

Although their goal was to better understand the role of heredity and environment in human development, Scott and Fuller nevertheless provided dog lovers with a wealth of information about how to understand behavior and nurture healthy development in puppies. Some things we've learned from Scott and Fuller: when puppies open their eyes (between 2 and 3 weeks of age, which is also when puppies begin moving backward in addition to forward); when they wean from their mother (between 7 and 10 weeks); the ideal age, from a human point of view, at which to take puppies from their mother if we want good pet dogs (10 to 12 weeks); the effects of social isolation (very bad; isolated puppies are either profoundly withdrawn and indifferent or overactive and hysterical).

One of the most important pieces of information—and one with tremendous implications for dog welfare—is that puppies go through a "sensitive period" between about 3 and 14 weeks of age, during which the stage is set for normal development of adult personality and behavior. During the sensitive period, puppies need to be socialized to other dogs, to humans, and to the various stimuli and situations they will encounter in life. Well-socialized puppies grow into confident, competent adults.

See also Domestication; Socialization

Selective breeding

Selective breeding refers to the deliberate intervention by humans in the reproductive output of dogs or other domesticated animals. Humans have been selectively breeding dogs—with varying levels of rigor—for thousands of years, increasing in precision and intensity in step with scientific understanding of heredity and genetics. Humans have selectively bred dogs for both morphological and behavioral traits. For much of our history with dogs, selective breeding targeted behavioral traits, particularly those relevant for dogs employed in various lines of work, such as guarding, herding, or pulling sleds. Over the past 200 years or so, selective breeding of dogs has come to be dominated by human aesthetic preferences. Selective breeding is responsible for the remarkable diversity in canine skull shape, tails, ears, coats, limb length, body shape, and body size. Selective breeding is easily one of the most consequential facets of human-dog relations, dramatically altering who dogs are and how we interrelate with them; it has arguably also become one of the most destructive forces in canine history.

See also Artificial selection; Digging; Landrace; Predatory sequence; Zhokhov Island

Shelters

An animal shelter is an establishment that provides a temporary home and care for pets who are straying, lost, rejected, or otherwise find themselves without a designated human home. Shelters also manage the killing of dogs who don't have a place in human society.

The term "shelter" was introduced in 1870s by Caroline Earle White, co-founder of the Pennsylvania Society for the Prevention of Cruelty to Animals, who thought the word "shelter" would be less open to ridicule than "home." At that point in history, busy urban centers in the U.S. and U.K. had large numbers of roving dogs. Attitudes toward these dogs were generally negative: people feared rabies and viewed mongrels as dirty and dangerous. The municipal response to loose dogs was to round them up and dispatch them, often by cudgeling, drowning, or hanging them on the spot. Disturbed by the very public and violent treatment of dogs, and inspired by the incipient humane movement, cities began to set up collection sites where dogs could be brought and where death could be administered with less pain and out of view. At the time, the killing of stray dogs was regarded as both necessary and an act of compassion. (This narrative is still largely in place.)

Although the original function of shelters was to kill dogs, the idea of rehoming dogs to new owners instead of killing them gradually gained traction. At first, it was primarily pedigreed dogs who were considered worth saving, but over time the goal expanded to include a wider range of dogs.

Shelters can be found in nearly every country but are more prevalent in places where dogs live almost exclusively within individual homes and where dogs cannot roam free. Indeed, the shelter is an essential component of dog-keeping practices in many countries. Shelters manage the population of dogs who have not been successfully integrated into a home, whether because of unexpected or unwanted behaviors, a change of mind, or an unforeseen change in circumstance. Shelters also support the dog industry, which capitalizes on the breeding and sale of live dogs: shelters allow breeders to continue to manufacture and sell puppies by absorbing excess stock.

From its humble beginnings, the shelter has propagated into an entire industry, with a market size in the U.S. of $3 billion, covering at least 12,000 shelters and 76,000 employees, not to mention an enormous army of volunteers. Roughly 3 million dogs enter a shelter each year in the U.S. Of these, around half will be rehomed. Some 670,000 dogs will be put to death within the shelter system, in a practice euphemistically called "euthanasia." And some will simply bide their time as residents of these canine halfway houses.

"Sheltering" often functions as a verb ("Here at Rover's Rescue, our passion is sheltering dogs"), but the word has also become popular in its gerund form. A word that describes an action (to offer physical shelter) has become a thing: the sheltering industry ("I work in sheltering").

See also Cynoctone; Pet; Sterilization; Stray

Small Dog Syndrome

A phenomenon observed by many a dog owner, and familiar also to dog behaviorists and veterinarians, is so-called Small Dog Syndrome. SDS refers to a loose collection of behavioral traits that are particularly associated with small dogs. Little dogs can be annoyingly yappy; they may engage in aggressive behaviors such as growling and nipping; may jump up on people; and may shake with fear at the sight of their own shadow. They are, in short, neurotic and ill-mannered. SDS is sometimes also called Canine Napoleon Complex, because these dogs can feel, to their human guardians, like short-statured household tyrants. Sometimes little dogs are blamed for being this way, but it isn't really their fault. It is usually the way they are treated by humans that makes them crazy.

SDS implies, by its name, that small dogs really want to be big dogs. SDS is sometimes taken to be a rough analog of small penis syndrome—a supposed lack of self-esteem and sense of powerlessness experienced by modestly endowed men, which leads them to buy obscenely expensive cars, boats, and watches. These two syndromes are totally unrelated. Small dogs don't likely have low self-esteem because of their small stature; a Chihuahua views himself as every bit an equal to a Great Dane. The problem is that small dogs haven't learned how to adapt to and live comfortably in human environments and may suffer from psychological distress because their motivational needs aren't being met. They may be suffering because they aren't really treated by humans as real dogs with canine needs.

Small dogs are often treated differently by their human caretakers than big dogs. They are less likely than their larger counterparts to get careful socialization and regular training. This is explained in part by the fact that aggressive small dogs are far less threatening than

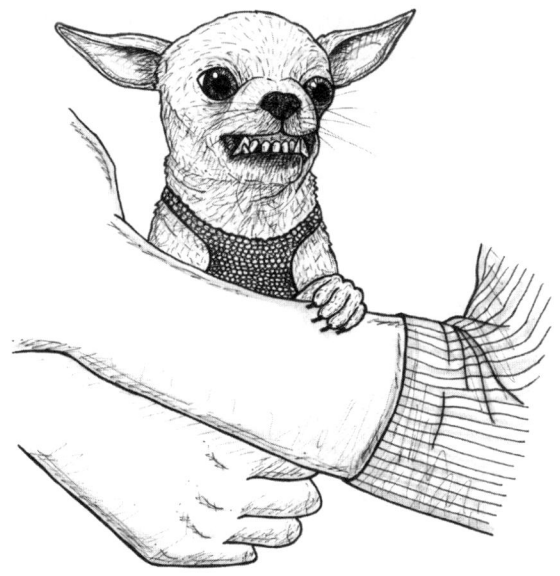

aggressive large dogs; they can easily be scooped up into their guardian's arms if they start going after a pedestrian's pants leg. When small dogs do get training, it is often in the form of cute tricks rather than life-relevant skills like recall and impulse control. Small dogs are more likely to be treated as toys or dolls, and their canine behavioral needs—for rough-and-tumble play with other dogs, for freedom to run around and scratch and sniff and roll in dead stuff—often go unmet. Although the label "Small Dog Syndrome" is used tongue-in-cheek, it reflects a serious and growing welfare problem. Currently, more than half of all dogs in the U.S. are small (under 25 pounds), and with a strong trend toward increasing miniaturization of breeds—including teacup and pocket varieties of already-small dogs—SDS could become even more widespread.

See also Queen Elizabeth pocket beagle; Yappy dogs

Sniffing

Sniffing is not the same as smelling. Olfaction, including smelling, is a form of sensory processing; sniffing is a behavior associated with olfaction. To put this another way, smelling is a non-cognitive process and occurs as an accompaniment to breathing. A dog cannot help but bring in smells. Sniffing, by contrast, is an effortful, purposeful behavior. Sniffing behavior is a way to receive olfactory signals, allowing a dog to scan the environment, localize odors, and discriminate between them. If you listen to a dog while he is actively sniffing something, you'll notice that sniffing is a rhythmic process, with the rhythm changing all the time, depending on context. You'll also notice, if you pay attention, that some sniffs are hard and fast: these fast sniffs enable quick odor recognition. Longer, slower sniffs increase the amplitude of the odor signal, but delay its transmission to the brain. You might notice a pattern of a long sniff followed by a burst of fast sniffs (or vice versa).

Dogs sniff all kinds of things, but one of their favorite things to sniff is each other. Whenever two dogs meet, they get busy sniffing. Sniffing is a mutual behavior— dogs do it to each other at the same time, together. Although canine sniffing behavior is not fully understood, some general patterns have been identified. Males initiate sniffing more often than females (no surprise there); males are more interested and sniff longer when females are in estrus (no surprise there, either). Males spend more time sniffing butts and genitals of females; females tend to be especially interested in sniffing the head. When dogs sniff butts and genitals, they are gathering information about who a dog is and how they are feeling. Sniffing can be a preliminary courtship behavior, and it can be a greeting behavior. Dogs also collect information about each other in absentia by sniffing urine, which leaves behind chemical signals.

See also Anal glands; Communication; Noses; Olfaction; Scent marking; Urine

Snuppy

Time magazine's 2005 Invention of the Year was a puppy. Born that year, Snuppy was the world's first successfully cloned dog. He was named in honor of Seoul National University (SNU + puppy), where the cloning research was conducted.

Animal cloning uses a scientific technique called *somatic cell nuclear transfer*. SCNT involves the transfer of the nucleus from a somatic cell into an egg cell (oocyte) that has had its nucleus removed. Somatic cells are any biological cells in a multicellular organism that are not reproductive—in other words, any cells other than sperm and egg cells. A sheep named Dolly, born in 1996, was the first cloned animal to live long enough to be labeled a "success." After Dolly came cloned cats, horses, mice, cows, rabbits, rats, pigs, and mules. It took scientists almost 10 additional years to clone a dog, though it wasn't for lack of trying. Dogs are more difficult to clone than some other mammals, in part because it is difficult to coax canine oocytes to grow in vitro (in a petri dish).

So instead, with Snuppy, scientists performed a nuclear transfer (transferring the nucleus) from adult skin cells into oocytes that had matured in vivo (inside a surrogate dog's body). The oocytes were collected from female donor dogs in the lab (breed/origin/name unknown); the skin cells were taken from the ear of a male Afghan hound. After fertilization, embryos were implanted into other female dogs in the lab who were used as surrogates. Snuppy's surrogate mother in the lab was a yellow lab. The second cloned dog, from the same experiment but a different surrogate mother, was labeled "NT-2" and died a week after birth. "Efficiency" of dog cloning at this time was very low: Snuppy and NT-2 were the only puppies born from 123 surrogates implanted with over a thousand embryos over the course of the cloning experiment.

At the time of writing, about 1500 dogs have been cloned at the South Korean university (at least, these are the reported numbers); about 20 percent of AKC's recognized breeds have been cloned, as have a number of mutts. Thousands of dogs have also been cloned by commercial companies. Exact figures are hard to come by, as pet cloning is not an industry known for its transparency.

Snuppy lived out his life at the SNU lab. He died at the age of 10 from cancer. Researchers don't know whether the cancer resulted from his cloning.

See also Working

Socialization

"Socialization" refers to the process of acquiring the social skills that will allow an animal to engage in species-typical behaviors. All mammals—domesticated and wild and including humans—go through a socialization process. For dogs, active socialization begins at birth, but the most important time for socialization extends from around three to eight weeks of age and is often referred to as the "sensitive period for socialization." What happens during this time greatly influences how dogs will interact with people, other dogs, and their environment. Humans raising dogs as pets as encouraged to expose puppies to a wide range of positive experiences and to be extremely careful to protect puppies from negative experiences such as being attacked by a dog or physically punished by a human, which may leave lasting psychological scars. One of the outcomes of "proper" socialization is a dog who is flexible, calm, resilient, and psychologically and emotionally well-adjusted.

See also Scott and Fuller

Solicitation behavior

Throughout the animal world, offspring engage in distinctive behaviors—sometimes called begging—that lead to provisioning by parents. The classic example is bird nestlings opening wide their helpless little beaks, in response to which parents drop in a bug or regurgitate some half-digested delicacy. Dogs are doubly smart, in that they not only use solicitation behavior to elicit caregiving from their canine mothers but have also figured out that they can use the same behavioral repertoire to solicit care from us, and not just when they are puppies.

Dog solicitation behaviors include whining, tail wagging (of a certain sort), yelping, licking (e.g., licking the face or hands of a person), pawing, and puppy-dog-eyeing. In response, they would please like to have whatever it is that you are eating. And your love, please and thank you.

See also Begging; Paws; Puppy-dog eyes; Wagging

Sterilization

Also often referred to as "spay/neuter" or "neutering," sterilization is the practice of surgically altering dogs so they cannot reproduce. In male dogs, this involves removing the testes; in female dogs, it involves either removing the ovaries, fallopian tubes, and uterus or just removing the ovaries. Surgical dog sterilization techniques were developed and refined with the explicit goal of trying to control the number of loose dogs on the streets. Sterilization was seen as an ethically superior alternative to killing. In 1972, the ASPCA made sterilization mandatory for all animals adopted from its network of shelters. It remains mandatory today and is often credited with reducing shelter euthanasia rates, although not all scholars agree that the purported relationship between higher sterilization and lower kill rates is so clear.

Although the main human agenda in sterilization programs is still population control, mass sterilization of dogs serves several other purposes. It serves the dog industry by restricting breeding to the dogs owned by breeders, which gives breeders a monopoly on puppy sales. Widespread sterilization also allows breeders to maintain purity of breed lines, which in turn allows humans to pursue their dreams of world domination at

Westminster. More recently, sterilization has been gaining traction as a form of behavioral modification, based on mostly folk beliefs such as "intact male dogs are more aggressive than neutered dogs" (for which claim there is zero good evidence). Finally, sterilization is sometimes motivated by health concerns about individual dogs.

Sterilization of dogs is a practice that varies considerably from country to country. In the U.S., sterilization is viewed as an essential task of the responsible dog owner. In other places, such as Norway, sterilization of dogs is considered cruel and is prohibited by law, unless medically indicated.

See also Reproduction; Shelters

Stray

Stray, as a noun, refers to any domestic animal—cat, dog, pig, cow—who has wandered off (accidentally, to be sure) from his or her human encampment. A stray has, by definition, gone awry, lost his or her way. In relation to dogs, the designation "stray" is used loosely, sometimes to refer to a pet dog who temporarily has no home, or sometimes to refer to a dog who is free-roaming, whose movement is unrestricted by a particular human, and who lives independently or semi-independently of humans. The term is problematic for reasons beyond its ambiguity. "Stray" carries negative and largely inaccurate connotations of dogs who, without a human owner close by, are dirty, hungry, aggressive, possibly even "feral" (another word with deeply negative overtones). "Stray" makes sense only in relation to "homed," and thus suggests that roaming free is an unnatural or undesirable way for dogs to live. It would help scientists studying dogs to have more objective, less ambiguous terminology.

A related designation is "loose." Like "stray," "loose" is imprecise, shifting, and highly ambiguous. Dogs are generally only called "loose" within cultures that have normalized the position of dogs as "kept." So, dogs get "loose" only in cultures where pet-keeping involves intensive confinement of dogs within an identified domestic home, and where dogs' movement outside of this sphere is tightly restricted. Outside movement must take place only in the presence of the dog's "owner," and only under constraint of a collar and leash. In the U.S. right now, being "loose" is both illegal for a dog and a rather tenuous situation. A pet dog can temporarily be loose, having discovered a door or gate that has carelessly been left open. But the assumption is that she will eventually be reunited with her family. Being "stray" tends to be a more permanent way of life. In cultures where dogs are "kept," "stray" is a far more degenerate condition than "loose." A stray has no permanent home, no human counterpart, and threatens the social order.

See also Cynoctone; Feral; Free-roaming; Pet; Purebred; Shelters; Sterilization; Streeties

Streeties

In some parts of the world, free-roaming dogs who live in cities are referred to as street dogs or, as a term of endearment, streeties. Street dogs can be found in virtually all cities, unless local human ordinances or attitudes are especially "anti-dog" and concerted efforts are made to exclude dogs from public spaces. But streeties are far more common in some places than others. India, for example, has a large and growing population of street dogs, due at least in part to steep population declines in vultures, who have long played a vital role in cleaning up carcasses of dead animals. Romania, Russia, Sri Lanka, and Turkey also have large numbers of streeties.

Street dogs may be dogs who were previously pets or dogs who have always been feral. Their attitudes toward and interactions with humans vary widely, with sliding scales of dependence on, comfort in the presence of, and friendliness toward people. Human attitudes toward street dogs also vary widely, both between cultures and even within cultures. Some people love street dogs, consider them part of the community, and support them with targeted feeding and affection. Others fear or hate street dogs and want them dead.

These dogs successfully exploit the ecological niche of the urban environment, feeding on human garbage and feces, preying on small animals that also exploit the urban environment such as rats and mice, and soliciting food from their human friends. Street dogs play an important cultural and ecological role in some urban environments and can have a positive impact on public health. They control urban animals who can carry disease and eat garbage and human waste, which can also transmit disease. At the same time, confoundingly, they play a negative role by being a significant vector for rabies.

There are very few to no streeties in the U.S., because we have created strict boundaries between pet and stray and really have no category for such dogs. A dog on the street would be labeled as loose or stray and would be captured and taken to a shelter.

The behavior of street dogs and other free-roaming dogs has become, in the past decade or two, a serious subject of scientific interest—finally! The behavior and ecology of urban free-roaming dogs appears to be different from that of rural free-roaming dogs, and both are different from dogs who live as pets. Unlike their wolf ancestors, street dogs rarely form packs—although they can and sometimes do. Most often, they live alone or in small groupings. They appear to live in fluid social structures. The behavior of street dogs almost certainly varies from city to city, country to country, depending on multiple factors (climate, food availability, competition with wild animals, attitudes of humans).

See also Free-roaming; Rabies

Submissive behavior

Submissive behaviors allow animals to maintain socially cohesive groups and reduce the high energetic costs and potential risks of aggression. Dogs use a variety of submissive postural, vocal, and perhaps also olfactory signals to avoid conflict. Some submissive behaviors you may have observed: lowered ears, lowered tail, crouched or cringing posture, lips drawn back, eyes averted or made small, peeing a little bit.

See also Aggression; Appeasement; Dominance; Yawning

Suckling

This tidbit may come as a surprise to some readers: there are numerous ethnographic accounts of women breastfeeding dog pups.

Tails

Tails

Most animals with a backbone also have a visible tail. (Poor humans have no tail, but only a "tail bone," which serves mainly as a target for potentially painful bruising after falling on ice.) Tails serve a variety of purposes for tailed, or *caudate*, animals; in dogs, the tail is especially important in communication and movement.

Tails are an essential part of dogs' communication toolbox. Tails add information to body postures and help dogs communicate mood and intention. For example, a tail held high usually projects confidence or arousal, while a tail held low is often a submissive posture. The interpretation of tail language requires attention to the broader context. A wagging tail can signal friendliness, nervousness, ambivalence, dominance, submission, and many other things as well, depending on how exactly the tail is wagging and what the rest of the dog's body—the face, ears, eyes, neck, hair, etc.—is trying to say.

The tail is an extension of the spinal column, and functions to help dogs keep their balance and maneuver through their environment. Tails come in especially handy when dogs are making tight turns with their body. (Watch a dog turning in a circle and observe how she uses her tail to help.) Dogs bred to be tailless or who have tails that don't function very well (super curly, super stubby) must work a bit harder to balance and lose out on a range of communicative possibilities. So do dogs whose tails are cut off on purpose by humans, a cruel practice known as tail docking.

In the world of dog breeding, breed standardization, showing, and judging, over 40 different types of tail are identified, many with amusing names and shapes.

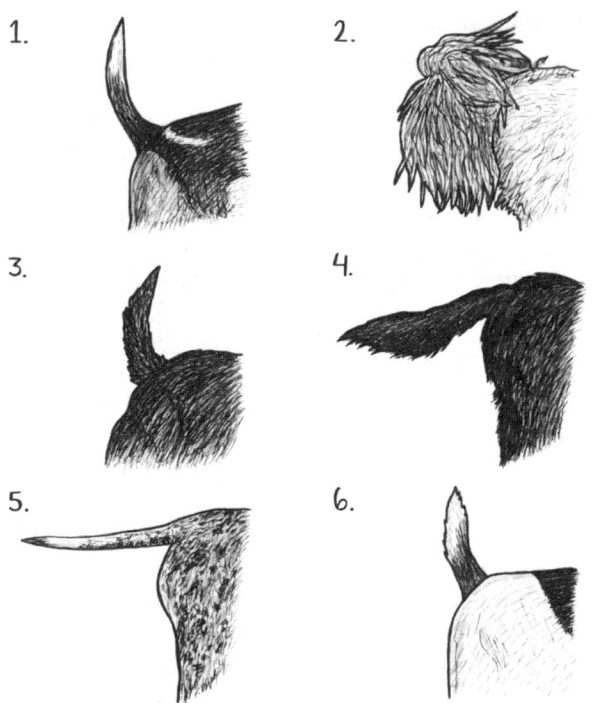

See how well you can connect tail type to breed:

A. bee sting   1. beagle
B. carrot-shaped   2. Pekingese
C. flagpole   3. Scottish terrier
D. otter   4. Labrador retriever
E. pipe stopper   5. German shorthaired pointer
F. squirrel   6. fox terrier

Answers: A. 5; B. 3; C. 1; D. 4; E. 6; F. 2

See also Communication; Lateralization; Wagging

Tapetum lucidum

If you have driven through the countryside at night, you may have caught the eerie glow of eyes hovering above ground. Many animals who are active in low light have a layer of tissue just behind the retina that reflects visible light back through the retina, brightening their visual world. This anatomical feature, called the *tapetum lucidum*, is what makes an animal's eyes shine green or yellow when caught in a car's headlights. The presence of the tapetum lucidum in dogs' eyes gives us some clues about their natural activity rhythms: dogs, like their canid cousins, can see well during the day and at dawn and dusk, allowing them a broad range of possible times to hunt and engage in other feeding activities. The tapetum lucidum is missing in some toy dog breeds, which is just weird.

See also Vision

Taxonomic classification

King Phillip Came Over for Good Soup. Many high school biology students will have this mnemonic for the Linnaean taxonomic classification seared into their brain. Using King Phillip as our guide: Dogs are animals (**K**ingdom: Animalia) with a supporting spine (**P**hylum: Chordata, subphylum Vertebrata), who are warm-blooded and produce milk for their young (**C**lass: Mammalia). Dogs eat meat as their primary source of food (**O**rder: Carnivora) and are members of a morphologically diverse **F**amily of dog-like carnivores called the Canidae. The Canidae have strong necks, long, well-developed limbs, five toes (one of which, called the dewclaw, doesn't reach the ground), pads under the paws, and prominent claws. The **G**enus *Canis* includes wolves, coyotes, jackals, and dogs, the latter being the Species *C. familiaris*.

Members of the family Canidae are called canids. There are currently 35 (or 34, or 36, or 37, depending on who you ask) extant species of canids in the world. Canids are one of the most geographically widespread of the carnivore families. Many of the wild canid species face extinction, although the domestic dog is not one of these. Indeed, domestic dogs may be driving losses of other canid species through introduction of diseases, competition for and disturbance of habitat, the human preference for domestic dogs and persecution of wild canids (e.g., coyotes), and hybridization.

Swedish botanist and zoologist Carl Linnaeus's binomial classification system assigns each species a two-word name (the so-called binominal nomenclature: Genus/species). In the case of dogs, Linnaeus used the Latin word *Canis* ("dog") for the genus; under this genus he listed *Canis familiaris*, the domestic dog, and *Canis lupus*, the wolf. In other words, dogs and wolves were considered separate species. With the development of DNA sequencing, scientists have been trying to further clarify the evolutionary relationships between dogs and wolves. Although there is still disagreement, and new genetic data are continually adding to (and complicating!) the picture, dogs are now widely considered a subspecies of the gray wolf. For the time being, then, their full Linnaean name is *Canis lupus familiaris*, but they are often referred to simply as *Canis familiaris*.

People wonder whether feral dogs, wild dogs, dingoes, and New Guinea singing dogs are the same species as domestic dogs. Feral dogs are most definitely domestic dogs, *Canis familiaris*. Wild dogs are most definitely not. The gorgeous African wild dog (*Lycaon pictus*) is a member of the genus *Lycaon*—and incidentally, the only extant, or living, member of this genus. The genus *Lycaon* is distinguished from *Canis* by the shape of its skull and teeth, the lack of dewclaws, and its hypercarnivorous diet (in which more than 70 percent of the food consumed is meat). The proper classification of the dingo, an ancient lineage of dog found in Australia, is still debated. The dingo is sometimes classified as a form of domestic dog (*Canis lupus familiaris*), sometimes as a subspecies of dog (*Canis familiaris dingo*), a subspecies of wolf (*Canis lupus dingo*), and sometimes as separate species altogether (*Canis dingo*). The classification of the New Guinea singing dog is similarly fraught and is mixed into the debate about dingoes. The NGSD is genetically similar to the dingo, but is it similar enough to be considered the same species, or does it represent a distinct population? You might ask whether it really matters how we classify. Well, it matters to scientists, who want to get things right. But it can also matter for the animals themselves, because classifications drive conservation decisions and management. Being designated as *Canis familiaris* can make conservation efforts more difficult. The domestic dog is not a threatened or endangered species. New Guinea singing dogs and dingoes are both threatened, partly by habitat loss and partly by hybridization with other canines.

See also Domestication; Hybrids; Wolves

Teeth

Teeth are one of the key features that make dogs identifiable as dogs, as opposed to another species of canid. Indeed, archaeologists studying the evolution of dogs measure the shape of skull and teeth to help decide whether a set of fossil remains was a wolf or dog or something in between. Dogs' teeth are also a good clue that they are carnivores. All four kinds of teeth in a dog's mouth are evolved for eating meat: the row of 12 little teeth in the front, called incisors, are used to scrape meat from bone (and useful, also, in removal of ticks and fleas and burrs, and to bite and nip people and other dogs); the 4 canines ("fangs") will puncture and hold flesh; the 16 premolars are used for shearing; and the 10 molars, way in the back, can crush and grind and chew. Even though dogs' teeth are shaped for carnivory, they can and do eat a whole variety of plant-based foods. Teeth also function in communication. Bared teeth are generally a threat, and the more teeth you can see, the higher the threat level.

The Tooth Fairy likes dogs better than humans. Humans only have 32 teeth, while dogs are endowed with 42 (and, oddly, the chow chow has 44 teeth). Like humans and most other mammal species, dogs are *diphyodonts*: they get two sets of teeth during their lifetime, a baby set and an adult set. In contrast, many other vertebrates are polyphyodonts, meaning the teeth are continually being replaced throughout the animals' lifetime.

The so-called milk teeth come in at about two weeks after birth. Because the puppy's mouth is small, he must start life with only 28 teeth. These deciduous teeth reach their full (small) size when a pup is between 8 and 10 weeks old, around the same time as mother dogs are starting to think about weaning (which makes sense . . . ouch). Puppies need very sharp teeth to chew up their food, because at that young age their jaws are

not very strong. Humans who choose to play with puppies should be prepared for a row of tiny razors to be
clamped onto loose fingers, noses, and other body parts.
The baby teeth begin to fall out at around 4 months of
age, and most dogs will have their full set of adult teeth
by the age of 6 or 7 months.

For pet dogs, dental care is a huge thing. At least in
part because of the way pet dogs are fed—more carbohydrates than their wild cousins; fewer opportunities to
rub off plaque while gnawing on stuff—they are prone
to dental problems. As with humans, if the saliva coat-
ing the teeth is not brushed off, it will develop into
plaque, which traps bacteria and leads to gum disease.
Most pet dogs will have developed periodontal disease
by the time they are several years old. Periodontal dis-

ease is one of the top three health threats to pet dogs (the other two are obesity and ear infections). Gum disease contributes to chronic and acute diseases, including heart disease, and accounts for significant (though as yet unquantified) losses in canine Quality-Adjusted Life Years. The good news is that puppies acquired as companion animals can learn, through careful training, to enjoy or at least tolerate having their teeth brushed.

See also Chewing; Communication; Dogor; Sagittal crest

Territoriality

How territorial are dogs? Is pee-marking on the fire hydrant an olfactory declaration of "This is MINE!"? And how many times have you heard a person say, as they wait for their dog to pee, "Hold on! He's marking his territory!"? Territorial displays are, indeed, part of the behavioral repertoire of dogs. But which behaviors are actually "territorial," and in which contexts, isn't always obvious, and how territory functions for domestic dogs is not well studied or understood. Dogs live in incredibly diverse spatial relationships to humans and to each other, making it very difficult to generalize about what behaviors might or might not be useful for dogs, and for confined pet dogs versus free-roaming dogs who are solitary versus free-roaming dogs who live in a pack versus all the other iterations you can think of.

Territory, within the field of zoology, is defined as the area an animal defends against competitors. Territoriality is the suite of behaviors by which an animal determines and defends this territory. Territorial behaviors are often agonistic (associated with conflict) and are sometimes even aggressive. Even so, the purpose of territorial behavior is to avoid having to fight, if possible. So, dogs have various behavioral displays meant to communicate boundaries, including urine scent-marking and ground scratching as well as vocalizations like barking. Dogs also use ritualized aggression—stiff body postures, raised hackles, bared teeth—to scare off potential threats. Territorial aggression may have been enhanced in dog breeds traditionally used for guarding or patrolling.

Resource guarding (e.g., a dog growling when you try to take away his bone) is not equivalent to territoriality, although both behaviors are motivated by a dog's need to maintain control over valuable resources.

See also Communication; Ground scratching; Scent marking; Urine

Tracking

Dogs are known for their tracking behavior, and their ability to track has been a key reason we've wanted to collaborate with them in hunting and why dogs are so useful in detection work. Tracking is olfaction at work. There are three phases of tracking behavior: searching (in which the dog tries to pick up the scent of whatever/whomever is being tracked); deciding (determining which direction the track goes); and tracking (following the track). According to one study of canine tracking, it takes a dog only five footsteps, on average, to figure out which direction an odor trail is going. That's amazing.

Dogs employ two basic modes of sniffing while tracking: ground sniffing, in which the dog keeps nose pinned to the ground, registering odor molecules and ignoring other olfactory stimuli; and air sniffing, in which the dog raises nose to the air, to pick up airborne molecules where no ground odors can be detected. Tracking is affected by weather conditions such as wind, humidity, and temperature. Although we generally associate tracking with noses and sniffing, it is likely that dogs are also using visual and acoustic stimuli.

See also Noses; Olfaction; Sniffing; Working

Tricolor coats

The diversity of coat colors among dogs, including the distinctive tricolor markings—black, tan, and white—seen on breeds such as the Bernese mountain dog and the border collie, has been the subject of surprisingly robust scientific debate. Why do dogs come in so many different color and marking patterns, while wolf coat coloring is . . . well, sort of boring? According to one story, coat colors have resulted from very careful, deliberate selective breeding by humans. We said, "Let's make dogs with white feet, black bodies, and tan eyebrows." And voilà! That's not a particularly compelling scientific explanation, for a variety of reasons. Another explanation for diverse coat coloration patterns is "domestication syndrome," the hypothesis (grossly oversimplified here) that selection for decreased aggression was linked, genetically, with a suite of other phenotypic changes, including smaller size, floppy ears, spotted fur, and shorter muzzles, among other things. This "syndrome" is evident not only in dogs, but also in cats, horses, rabbits, and a few other domesticated species. Scientists are still arguing about whether domestication syndrome is really a thing and, if so, how we should understand it—and about whether humans also have domestication syndrome.

Meanwhile, the question of coat coloration continues to get more complicated. In 2021, geneticists found evidence that diverse coat patterns were around at least 10,000 years ago, before any selective breeding, and perhaps much earlier, even before dogs became dogs. Researchers looked at 5,000-year-old DNA of ancient dogs from Eurasia and found regulators of the agouti-signaling protein (ASIP) gene. The ASIP gene controls the expression of yellow pigmentation, which can appear tan or white. Agouti is a fur coloration pattern in which each hair displays two or more bands of pigmentation. It is seen in various animals, including rabbits, horses, cats, mice, and the eponymous agouti (genus *Dasyprocta*), a species of large rodent native to Central and South American rainforests.

The DNA of one fossilized dog—the Zhohkov Island dog from Siberia, dated at around 10,000 years old—already showed what is called the "black back pattern," which looks suspiciously like the coloring of a Bernese mountain dog.

See also Domestication; Zhokhov Island

Tripods

Tripod (or sometimes "tripawd") is the colloquial name for dogs (usually pet dogs) who have three legs. These dogs have often lost a foreleg or back leg to traumatic injury or amputation. Tripods need special care, such as elevated food and water dishes, non-slip flooring, and vigilant attention to maintaining a healthy weight. Yet with simple modifications to their care, tripods can live joyful, fulfilled lives. Tripods have a devoted human following, including Facebook pages, support

groups, and information platforms. The message that tripod lovers want to get across is that dogs can readily adapt to challenges, including life on three legs—even, in some cases, two legs. Tripods bring attention to the breadth of human ingenuity in caring for dogs and other companion animals. Although a three- or possibly even two-legged dog could conceivably survive without human assistance, human caregiving and ingenuity certainly make adaptation far more likely and comfortable. People have come up with ingenious ways to make life better for animals with disabilities. A wide range of prosthetics, wheelchairs, strollers, and braces have been designed to help dogs stay mobile and keep engaged with the world. It could be said that dogs have contributed a disproportionate quantity of care and kindness to the human-dog partnership, but tripods remind us of how much humans are willing to do for dogs.

See also Love

U nderdog

One of the earliest recorded uses of the English idiom "underdog," the dog beaten in a fight, was in 1887. "Top dog" appears to have come into use several decades later. "Underdog" referred to a person or team or country not expected to win in a fight, especially used in sport and other competition. The funny thing about the underdog is that nearly everyone is rooting for him or her. What's not at all amusing is that these idioms have roots in the nasty world of dog fighting.

Dogs find their way into all kinds of idiomatic expressions. Just as they are sprinkled through our lives, they are sprinkled into our language. One sees an unfortunate pattern of dog-related terms used in deprecatory fashion in the human realm: the word "dog" has itself long functioned as an insult, to tell someone they are worthless. When applied to a woman, the insult takes on the added injury of "ugly," though the term may be used playfully to refer to a rakish man ("you dog!"). Other dog-sourced insults: "Bitch": a woman whose behavior is transgressive; "churlish" (derived from "cur"): the temperament of a mean person. Dog idioms also reflect the hard-working life of dogs in times past: "work like a dog"; "be treated like a dog"; "a dog's life"; "go to the dogs." "In the doghouse" refers to someone who is in trouble for having misbehaved. "Hair of the dog [that bit you]" refers to the rather questionable advice given to someone with a hangover to have another drink. Other common expressions that we may not even readily identify as dog-sourced include "watchdog," as in governmental oversight organizations; "son of a bitch"; and "bone of contention." The word "sic," as in "sic 'em," is a modification of *seek* and was originally used (circa 1845) as a command to set a hunting dog on her quarry or a guard dog on an intruder. "Abet" comes to us from the medieval sport of bear baiting and is the English form of Old French *abeter*, "to hound on," to urge a dog to attack.

Dog idioms are also found in politics. A "dog whistle" is a subtly aimed political message which is intended for, and can only be understood by, a particular group. A "yellow-dog contract" was an agreement made between an employee and employer, sometimes as a condition of employment, that the employee would not belong to a labor union. Yellow-dog clauses were outlawed in 1932 in the U.S. The exact origin of the term "yellow dog" is uncertain, but it was most certainly not a compliment. It may have been a reference to mongrel dogs, who at that time were often yellow in coat color; mongrels were generally considered worthless and dirty. The term "yellow dog Democrats" has been applied, since the late nineteenth century, to rigid party-line Democratic voters, who would allegedly vote for a yellow dog before voting for any Republican. A so-called political attack dog is a verbal mercenary who will mercilessly attack a political opponent.

Nearly all dog idioms and phrases paint dogs in a negative light or remind us of horrible things that humans have done to dogs. But there are some exceptions. A book with "dog-eared" pages—pages turned down on the corner, presumably to mark a reader's place or to mark an exceptional passage—is worn and unkempt, but in a way that suggests a book so deeply loved as to be reread many times. The expression first appears in the mid-seventeenth century. Another lighthearted expression is "The dog ate my homework," which may have first started circulating as a schoolroom excuse in the early twentieth century. This idiom points to a broader human habit of using dogs as scapegoats. Someone might, having let escape a particularly stinky fart, turn to the dog and say, "Wallace! That's disgusting!" ("Scapegoat" is another animal idiom of interest and refers to a ritual described in the biblical book of Leviticus, where one of two kid goats was set free into the wilderness, having had the sins of the people laid symbolically upon his head. The other goat was sacrificed.)

Urine

Urine is a waste product of metabolism, formed in the kidneys, stored temporarily in the bladder, and excreted through the urethra during a process called urination (or micturition). Feces and sweat are also metabolic waste products. The excretory system—comprising the organs that remove metabolic waste from the body—works in concert with the digestive system to maintain homeostasis. Dog urine contains water, urea, creatinine, uric acid, enzymes, fatty acids, ammonia, magnesium, and hormones, among other things. (Fun fact: in mammals, urine comes in liquid form; for many birds and reptiles, urine is solid or semi-solid.)

Urine is not just a waste product, though. Many animals use urine as a communicative tool, and dogs have honed this skill. Urine is one way in which dogs "talk" to each other, both in real time ("Watch! I'm peeing now" as a visual display) and through olfactory messages that persist in the environment for an extended period. Urine allows dogs to communicate with one another about who was where and when, about potential reproductive availability, and perhaps also about mood and emotion. Peeing leaves messages for other dogs to read. Sniffing pee spots is how dogs read through these not-so-secret notes left behind by other dogs. Dogs are curious to smell the urine of other dogs, and they are motivated to pee on all sorts of things, including on top of the urine of other dogs, a behavior called overmarking. Although the function of overmarking is not fully understood by humans, dogs may mark to cover the scent of other dogs or to highlight their own scent. Ethologists use the charming phrase "competitive countermarking" to refer to overmarking of a single pee spot by various dogs, a behavior that can lead to "fire hydrant syndrome"—in which a landmark collects so much dog pee that it is noticeably stained, and even humans can smell the overstuffed olfactory mailbox.

Research on dog elimination behavior suggests that urination (and defecation, too) has a broad range of meanings for dogs. Some urinating may be territorial, but much of it is not. Sometimes dogs pee because they want to mask the odor of another dog's urine or be sure their scent is the one that others detect. And of course, sometimes dogs pee simply because their bladder is full.

Male dogs often lift a leg to scent mark with urine, and occasionally female dogs will also lift a leg. Little dogs can exaggerate their signal by lifting a leg very high, leaving the impression that they are bigger than they actually are (a clever way to "grow" themselves). Dogs and their wild relatives will occasionally lift a leg without depositing any noticeable urine, a behavior known as

"dry marking." It isn't clear exactly why dogs dry mark, but leg lifting may serve as a visual signal telling other dogs that pee was deposited, even if it wasn't. Pee is an important resource, and dry marking would allow dogs to make a statement while still conserving pee for later, when it might be even more important. Dogs will often dry mark and then, within a few seconds, lift a leg and pee, suggesting that dry marking isn't just a matter of a dog having "run out" of pee. Ethologist Marc Bekoff's research found that dogs dry mark more often when there are other dogs around who can see them, giving support to the "dry marking as visual display" hypothesis. Dogs also often scratch the ground after peeing, which may be a way to accentuate both the visual and olfactory elements of the communication.

The topic of dog pee is incomplete without mention of the fact that canine urination behavior is a flash point in human-dog relations. One of the first things many humans do with pet dogs is to shape elimination behaviors so that dogs pee only in prescribed places and on a time schedule that suits the needs of the humans in question. "Inappropriate" elimination—an unwillingness or inability to adapt to this prescribed schedule—is a primary reason that dogs are ejected from human households and sent off to shelters. Dog pee can also become a serious point of contention between human neighbors and within human communities. Because of its high nitrogen content, dog pee can create yellow patches on grass lawns; it can also alter soil microbial communities, making large quantities of dog pee a potential problem in sensitive ecosystems.

See also Communication; Ground scratching; Olfaction; Scent marking; Territoriality

Vision

It is often said that dogs have poor vision, at least relative to humans. But this isn't true—they just have different visual strengths than we do. Visual acuity in humans is often described using the Snellen fraction, which anyone who's been to an ophthalmologist will have heard mentioned. Perhaps you've been assured that you have 20/20 vision. Visual perception is hard to study, but as best researchers can tell, dogs are nearsighted. Using the Snellen chart, dogs have roughly 20/50 vision, meaning that what we see at 50 feet (assuming our vision is perfect), a dog would only be able to see from 20 feet away. (Famously sharp-eyed eagles have 20/5 vision.)

Experiments on dog color perception suggest that while not color-blind, dogs likely may not see as many colors or may not see all the same colors as humans. Dogs, for example, can distinguish blue from yellow, but not red from green. Research has found that the lenses of dogs' eyes transmit significant amounts of ultraviolet light. These wavelengths are invisible to people (unless they are missing an eye lens). Dogs' eyes are more sensitive to motion than ours, which helps explain why they take off after a squirrel well before we are even aware that there is a squirrel to chase. Dogs are visual generalists, so their eyes work well in a range of light levels. Because dogs and other canids tend to be especially active at dawn and dusk—they are *crepuscular*—their vision is good in low-light conditions.

See also Lenticular sclerosis; Tapetum lucidum

Wagging

In today's pet industry marketing extravaganza, almost nothing is more strongly associated with dogs than a wagging tail. And these wagging tails are almost universally taken to mean Dogs Love Us! A wagging tail signals that our dogs are happy to see us and are feeling good about their lives. And tail wagging is, often, a sign of friendliness and positive emotions. But not always. Wagging can signal uncertainty, ambivalence, and even threat. The messaging contained in a wag depends on its speed, direction, and height relative to the dog, among other things. An easy, horizontal tail wag typically indicates a friendly dog. A stiff, high wag might signal dominance. A low wag may be sign of submission, especially when combined with a few drops of pee.

A wag is like one word in a sentence. You need the whole sentence to figure out the meaning of the word in that specific context. When the word is "wag," some other parts of the sentence might include body posture (head, back), ears, eyes, mouth, fur, and vocalizations. If a wag is part of a sentence which includes a rigid body, raised hackles, and bared teeth, then it is safe to say this dog is not in the mood for a great big hug. If they have been well socialized, dogs are usually pretty good at interpreting each other's wags. Humans, not so much. Many dog bites occur because humans have misinterpreted a wagging tail.

See also Communication; Emotions; Lateralization; Tails

War

Toward the end of Act 3, Scene 1 of Shakespeare's *Julius Caesar*, Mark Antony is alone with Caesar's body, shortly after the dictator has been assassinated. As his soliloquy draws to a close, Mark Antony swears to Caesar's departed spirit that he will seek to avenge his death, that he will "Cry 'Havoc!' and let slip the dogs of war." These words are not just a metaphor for unleashing soldiers to commit all manner of raping and pillaging upon the enemy. Attack dogs really have been released in battle. And dogs have long worn slip collars, which tighten when a dog pulls but can be loosed or thrown off or "let slip," releasing dogs to charge forward and attack.

The battlefields of history are drenched in the blood of dogs. Dogs have been participants (willing or unwilling, it's hard to say) in human warfare across time and place. Dogs were present in the wars of the Romans, Egyptians, Greeks, Persians, Vikings, and British, to name just a few. They were present on battlefields of Europe during the Middle Ages and Renaissance, and they fought in both world wars. Dogs are still an active part of military training and maneuvers around the world.

As Shakespeare's text suggests, dogs were often sent into battle first, to incite terror and disrupt enemy lines. Dogs were both physical combatants and psychological weapons. Dogs have been valued not only for their fearsomeness, their brute physical strength and strong, flesh-tearing jaws. Dogs' keen sense of smell has been exploited: dogs have been used to detect land mines and explosives and to track the scent of retreating enemy soldiers. So, too, has their superior ability to hear, which has made them useful as sentries and guards, alerting soldiers to the approach of enemy combatants.

Dogs have experienced war in other ways, too. When towns and farms and cities were bombed and burned and people lost everything, dogs lost everything, too. The neighborhoods, homes, and families of pet and free-roaming dogs were fractured. Countless dogs died as civilian casualties. Images from the war in Afghanistan, for example, include U.S. soldiers shooting free-roaming dogs for amusement. Cruelty toward dogs is not unique to war, but war serves to accentuate violence to animals, by loosening the moral reluctance to kill. Dogs have been victims of cruelty, and victims of kindness. One devastating example occurred in London during the Second World War. In preparation for the German bombings that the English feared were inevitable, people were encouraged to bring their pets to the "pound" for preemptive euthanasia. Hundreds of thousands of dogs and cats were killed as an act of kindness, to prevent them from suffering the terror of and potential death from bombing.

Dogs are not the only species of animal used in war. Horses have played as big, if not a bigger role than dogs, used to accentuate the speed, force, and size of humans riding on their backs, and as "horsepower" to build fortresses, carry supplies, and transport soldiers. A vast menagerie of animals has been involved in human warfare in one way or another, including camels, pigeons, pigs, dolphins, elephants, and oxen. Some animals have been turned into weapons by having bombs attached to their bodies; some have been used for sending messages across enemy lines; some have served as mascots and mood lifters. The human imagination seems to know no bounds when it comes to warfare.

See also K-9; Literature; Weaponizing; Working

Weaponizing

Dogs have not only been asked to fill various roles during wartime but have also been employed as tools of colonization and oppression. For example, when the Spanish conquistadors came to the Americas in the sixteenth century, they brought with them enormous mastiffs who were trained to hunt down and disembowel fleeing villagers. Led by Columbus, the conquistadors who claimed American soil as their own also used these dogs to instill fear and inflict physical injury on native populations. Later conquistadors such as de Soto, Cortés, and Pizarro continued the tradition of using dogs to inflict genocide on indigenous populations. It wasn't that the native populations had never seen dogs before—they lived with and among their own village dogs; dogs were part of their culture, their mythology, their everyday experiences. But the dogs brought by the

Spanish were almost unrecognizable: they were enormous and were trained in brutality.

The establishment of slave plantations on the Canary Islands and in the Caribbean likewise involved the use of dogs to terrorize and subjugate native peoples. The use of dogs as weapons of enforcement in slave regimes continued in the American South, where dogs were used to enforce plantation work regimes and to hunt down any fugitive slaves trying to escape. Dogs were used not only to force labor and compliance through intimidation, but also for their tracking skills—in this case, they were used to track and hunt humans. Plantation owners bred bloodhounds imported from Cuba and Germany and trained the dogs to be ferocious toward Blacks. This "training" sometimes involved forcing enslaved people to beat and terrorize the dogs, to create negative associations. Fugitives came up with innovative ways to evade the slave hounds by covering their own scent with turpentine, onion, or the smell of rabbits.

Dogs were considered by the conquistadors as more intelligent and sentient than the native people. This ugly motif was repeated in the antebellum South. Southern "slave hounds" were often better fed and more highly regarded than the slaves they were bred to track and kill. Objections to the use of "slave hounds" helped motivate the abolitionist movement.

See also Racism; War; Working

Whiskers

Whiskers are considered a sensory organ, alongside the eyes, ears, and nose. The formal name for these modified hairs is *vibrissae* (from the Latin *vibrare*, "to vibrate"),

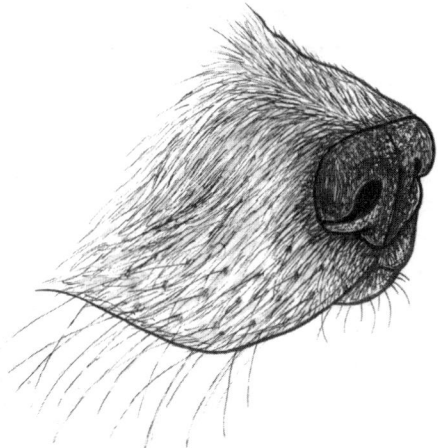

which gives you an important clue to their function. Whiskers help animals gather tactile information from the environment, and can aid in navigation, finding food, sensing danger (e.g., by detecting shifts in air currents), and communicating with others.

Most mammals, except for humans and a few monotremes such as the duck-billed platypus, have whiskers at some stage of their life (some just as youngsters). In many terrestrial mammals, including dogs, the whiskers are clustered into four main groups on the face: on their upper lips, lower lips and chin, above the eyes (where they might look like part of the eyebrow), and on their cheeks. Each whisker follicle has its own blood and nerve supply and is distinctly represented in the sensory cortex of the brain, making it extremely sensitive to touch.

It isn't entirely clear what function whiskers play in the lives of domestic dogs, especially pet dogs. They are probably used in social communication. For instance, the visual signal of pulling back a lip might be accentuated by whiskers, as might the raising of eyebrows. But scientists don't know for sure.

Like the rest of the hair on a dog's body, whiskers are made from keratin. But they are different from the pelage hair: whiskers are longer, stiffer, thicker, and more deeply embedded in the skin, and serve different functions. Pelage is the hairy, woolly, or furry coat found on almost all mammals (once again, humans come up lacking). It serves many different functions. The pelage insulates the body from heat or cold and protects an animal's skin from getting scratched, torn, or otherwise injured. Pelage coloring often matches the environmental background, assisting in camouflage. Pelage is also part of an animal's communication system (recall the discussion of hackles, which involve puffing out the coat to look threatening).

See also Communication; Hackles

Wolves

Modern-day dogs are descended from an ancient and now extinct ancestor. The gray wolf, which ranges across North America, Europe, and Asia, is the dogs' closest living relative. Dogs share with gray wolves about 98.8 percent of the same DNA, and ethologists estimate that dogs share perhaps a half to a third of the social behaviors of wolves, although some of these behaviors have been modified. We can learn a lot about who dogs are by studying wolves, and understanding a bit about wolves can help dog guardians provide their companions with meaningful lives by revealing some of the basic behavioral repertoire of canids. At the same time, the use of the wolf as an analog or guide for studying dog behavior and interacting with pet dogs has been, at times, profoundly misleading and damaging to dogs, most notably in the misunderstanding of the concepts of dominance and pack hierarchy.

See also Dominance; Sagittal crest; Taxonomic classification

Working

The work dogs do has changed over time and varies across geography. But one thing remains constant: humans rely on dogs for an incredibly diverse range of support, and we certainly wouldn't be who we are without the labor of dogs.

It is hard to categorize the work dogs do, because it is so diverse, so complex, and so multilayered. But we might roughly differentiate sensory, physical, and emotional labor. Sensory labor employs a range of canine capacities, including their sense of smell (their ability to detect odors, track a scent) and their sharp ears (as they listen for intruders). A good deal of canine work falls into the category of physical labor: dogs are used for their strength, speed, and size. And a growing category of canine labor could be categorized as emotional. Dogs function as companions, mood lifters, and stress relievers, both informally as family members and formally as therapy dogs. Most canine work combines two or three of these categories. For example, dogs working in search and rescue are doing sensory work (using their olfactory skills to find survivors), physical work (traveling over rough terrain in difficult conditions), and emotional work (providing emotional support to rescue workers).

Here is a list of some of the many kinds of work dogs do for and with us.

SEARCH AND RESCUE. Dogs take part in avalanche rescue; finding lost or injured hikers; disaster response and recovery (floods, earthquakes); 9/11 (dogs searched for survivors and for remains).

LAW ENFORCEMENT. Dogs help find hidden explosives, cadavers, illegal drugs, and counterfeit money. They are also used to help police control crowds and to chase and subdue suspects.

MILITARY. Dogs are used to find bombs, track soldiers, send messages, and scout for danger. Dogs also play an essential role in detecting landmines, which falls into the category of military labor, but also into the category of humanitarian work. Some 100 million landmines are scattered around the world, remnants of past human conflicts. Dogs can locate mines far more effectively and safely than people, though the work is risky.

MEDICAL DETECTION. Dogs' keen sense of smell can detect certain cancers through urine, sweat, breath, feces, and skin. Some of the cancers detected by dogs are melanoma, colorectal cancer, lung cancer, and breast cancer. Dogs can also detect certain viruses and bacteria (e.g., COVID-19). At this point, the medical detection skills of dogs are mainly a curiosity and have not been broadly applied within health care, nor is it necessarily feasible to employ dogs in this realm on a large scale.

THERAPY. Dogs are present in hospitals, nursing homes, airports, schools, and at disaster sites, where they support victims and rescue workers.

SERVICE. Dogs began being specifically trained to assist the blind in the 1750s. Since then, dogs have been enlisted in a wide variety of roles that fall under the label of "service animals." A service animal has been individually trained to perform tasks for people with disabilities. Dogs help people with visual and hearing impairments; provide mobility assistance; provide support for people suffering from PTSD; and have been used as an advance warning system for people who experience seizures, hypoglycemia, or narcolepsy.

CONSERVATION. Conservation is a relatively new and
exciting realm of canine work in which dogs assist
with efforts to conserve biodiversity. Dogs can effectively locate biological organisms of interest to conservation scientists. They have been employed to help
control invasive species. For example, dogs helped
slow the spread of brown tree snakes in Guam. Dogs
have been put to work finding palm weevils in date
palm crops, and in locating gypsy moths, termites,
and screw worm flies. Dogs can also detect cyano-
bacteria and other microorganisms.

HERDING. Dogs help manage sheep, cattle, and other
domesticated animals. Dogs keep a flock or herd of
animals in a group, help move them from one place to
another, and pick up stragglers. Other dogs are used
to protect the herd from wolves or other predators.

HUNTING. Dogs have been hunting collaborators since
the very early days of our co-evolution. Dogs help in a
variety of ways, from tracking prey, to alerting us to the
presence of prey, to chasing down and catching prey.

SLED PULLING. For thousands of years, dogs have been
put to work pulling things for humans. What may
come to mind first are dogs pulling sleds in the snow.
But dogs have also been put to work pulling carts and
other contrivances and were used to power turnspits
and cider churns.

GUARDING. Dogs work as sentries, listening and smelling for intruders. They also provide physical protection by means of bodily force. Dogs protect people
and property, including livestock.

EMOTIONAL SUPPORT. Dogs work as "emotional support animals," providing therapeutic benefit in the

way of comfort and companionship to people with psychiatric disabilities.

BIOLOGICAL SUBSTRATE. Although it is not typically acknowledged as a form of canine labor, dogs have been and still are used as biological models to study human disease and test human drugs. Dogs' bodies are also used to produce puppies for sale in the pet market.

PET-ING. Pet-ing (gerund form, to be distinguished from "petting," as a verb) should be added as an additional category of canine labor. Often a distinction is drawn between pet dogs and working dogs, based on the assumptions that pet dogs serve no useful function and that they lead lives of leisure. Both assumptions overlook the seriousness of dogs' work as companion animals. Being the companions of humans is not easy—we are demanding, emotionally labile, needy, and often cruel. Because dogs are empathic toward and emotionally synchronized with their human guardians, they absorb a lot of our stress and sadness—and this can take an emotional toll on them. Companion dogs often suffer from the excessive demands of emotional labor while at the same time being deprived of dog-relevant work, such as procuring their own food or negotiating complex social interactions with other dogs, which might provide a positive sense of accomplishment and mental well-being.

We tend to conceptualize the work of dogs in terms of labor or skills they provide for us. But dogs also work hard to fend for themselves and their families. Dogs who live semi-independently of humans, as free-roaming, street, village, and feral dogs, are also "working dogs."

Canine labor is ethically fraught and raises a range of hard questions. Is it fair to make dogs do dangerous work? Do they consent to their work in any meaningful way, and should consent be a required element of our collaborative work? Does providing dogs with food and shelter count as a reasonable form of pay? Should dogs be compensated in additional ways when they do risky work?

See also Conservation impact of dogs; K-9; Olfaction; Pet; Pet effect; War; Weaponizing

Xenophobia

Do dogs recognize others of their own breed? Do they like them better than dogs of different breeds or of mixed heritage? Do dogs ever exhibit prejudice against breeds that aren't like them? According to loads of anecdotal evidence from dog guardians, a canine version of xenophobia does, in fact, seem to occur. For instance, Poppy, a self-contained and dignified herding-type dog, seems to dislike all curly-haired dogs, especially poodles and goldendoodles. On the flip side, many dog guardians report that their dog seems to recognize and get especially excited when meeting others of their own breed. Ody the vizsla cannot contain his excitement if he spots a fellow vizsla at the park. He seems to recognize other vizslas from hundreds of feet away.

But stories from dog guardians are all we have so far. No one has figured out how to study these seeming canine preferences and prejudices, and as far as the science goes, there is nothing to indicate that dogs recognize and prefer others who look just like them. When it comes to in-breed loyalty, it is possible that dogs are using some chemosensory clue that indicates genetic relatedness. As for xenophobia, it may be that dogs form impressions based on the general shape and size and other physical characteristics of dogs, and once offended by a golden-doodle, always offended by all goldendoodles.

Taken as an innate fear of strangers—unfamiliar dogs and unfamiliar people—xenophobia is common across all dogs, and indeed, across all canids. Being cautious around unfamiliar others is a useful evolutionary strategy for animals who live in social groups.

Xenophobia, of course, is a well-established human behavioral trait. What's interesting for our purposes is how often human prejudice gets channeled through attitudes toward dogs. For example, in his fine history of dogs, Chris Pearson notes that in the late nineteenth and early twentieth centuries, when pedigreed dogs had been firmly established as symbols of class membership and national pride, the wrong kinds of people were associated, naturally, with the wrong kinds of dogs. One British commentator noted that his homeland would be better off without chow chows, "weird" Alsatians, French police dogs, and other "freaks imported . . . from the dark places of the earth."

See also Racism

Xolo dog

The Xoloitzcuintli (pronounced "show-low-eats-QUEENT-lee"), affectionately known as the Xolo dog, is one of the oldest breed lineages. The name derives from the Aztec words *Xolotl* (god of lightning and death) and *itzcuintli* (dog). In Aztec mythology, Xolotl made the dog from a sliver of the Bone of Life, as a gift for humans.

Humans were instructed to protect the Xoloitzcuintli. In return, the Xolo would guide humans through the underworld after death. To this end, Xolos were often sacrificed during Aztec funerals. They were also a common source of meat.

Xolos are hairless (although sometimes puppies are born with fur)—a useful adaptation to life in hot and humid environments. Indeed, another name for the Xolo is the Mexican hairless dog. Their skin has a bluish/grayish tint, and on some Xolos a curious tuft of hair runs down the head. Although Xolos are often entered by their owners into World's Ugliest Dogs contests, the dogs strenuously object to this indignity.

See also Landrace

Y appy dogs

"Yappy dog" is a derisive term used to refer to the class of small dogs, frequently kept as fashion ornaments, who have high-pitched barks and who bark more than is strictly necessary. Little dogs have higher-pitched and, to some human ears, more annoying barks, than big dogs, because little dogs' vocal cords are shorter and smaller—so they really can't help it if they grate on our nerves. Unfortunately, many people don't take yapping seriously as a legitimate form of canine communication. Humans have a prejudice against small barkers: little dogs yap (irritating, meaningless noise) while big dogs bark (a dignified form of communication). Little-dog barks need some respect.

See also Barking; Communication; Small Dog Syndrome

Yawning

Dogs sometimes yawn in response to another dog yawning, an example of what is called emotional contagion. Dogs may yawn when feeling stressed or when experiencing pain. A yawn can be a calming signal, used to lower the level of arousal in another dog, or, alternatively, it can be a form of self-soothing. Yawning can also be a sign of insecurity, ambivalence, or motivational conflict. A yawn can be a displacement behavior. A dominant dog may yawn to show friendliness; a submissive dog may yawn to signal that he or she doesn't pose a threat. And sometimes, of course, a yawn is just a yawn. Your yawning dog may be feeling sleepy and ready for an afternoon nap.

See also Appeasement; Displacement behavior; Emotions

Yellow snow

Until just a few decades ago, the consensus among philosophers and scientists was that humans possess a range of capacities that evolution bequeathed to us alone among all beings on the planet—a ludicrous proposition, biologically speaking, but this hasn't stopped anyone from believing it. Some of the capacities thought to be unique included tool use, culture, emotion, mourning, and morality, all of which we now know to be broadly distributed among Earth's taxa. Self-awareness, or an awareness of the self as an individual distinct from one's environment, is another capacity long believed to be unique to humans. Yet it appears once again that we are going to have to share, all thanks to some yellow snow.

Despite protracted disagreement about how to define self-awareness—sometimes, confusingly, also referred to as self-recognition, self-consciousness, and/or having a "theory of mind"—scientists have been busily engaged in determining which animals have it and which don't. So far, the number of species "proved" capable of self-awareness is narrow (with "proved" here in scare quotes because, as you'll see in a moment, our capacity to imagine and investigate the minds of other creatures is limited by—of all things—our own minds). Humans, obviously, are well-endowed with self-awareness. Also, apparently, chimpanzees, elephants, dolphins, magpies, and a fish called the cleaner wrasse. These species have all passed what is known as the mirror test, the current gold standard for determining whether an animal recognizes itself as a Self. In the mirror test, an unsuspecting animal is sedated and, while asleep, adorned with a spot of red paint on the forehead. Once awake, the animal is then placed in front of a mirror. If the animal is perplexed by the red spot and tries to rub it off, they are deemed to have self-awareness.

Dogs have repeatedly failed the mirror test. Poor things. Yet our intuition screams at us that dogs are self-conscious, self-aware, self-recognizing beings. Could the test itself be problematic? Some scientists think so. Perhaps the fact that the test relies exclusively on visual information biases the test toward species, like humans, whose evolution has favored visual processing. Recognizing the limitations of the mirror test in relation to dogs, whose interface with the world is heavily mediated through olfaction, scientists have looked for alternatives. Ethologist Marc Bekoff, for example, suggested that we might explore self-recognition using scent. Using his own dog Jethro as test subject, he scooped up small piles of snow where Jethro had peed and, without Jethro's knowledge, transplanted the urine-infused snow to new locations. He then walked with Jethro and recorded the amount of time Jethro spent examining his own urine spots compared to those of other dogs. The basic hypothesis was that if Jethro recognized his own scent, he wouldn't need to spend time sorting out the olfactory messages left in the snow. He would just say, "Oh, that's me," and move on. Jethro did, in fact, spend less time examining his own urine spots than those of other dogs. Since Bekoff's original Yellow Snow test, scientists have set up larger sniff self-recognition tests for dogs and this on-going research will continue to expand our human understanding of canine minds.

See also Olfaction; Urine

Zhokhov Island

Early human inhabitants of Zhokhov, an island in the frigid East Siberian Sea, may have been among the very first to selectively breed dogs for a particular purpose. On snowbound Zhokhov, that purpose was pulling sleds. In 2017, scientists published findings from their analysis of archaeological artifacts found on Zhokhov Island, including fossilized dog bones and the remains of wooden sleds. They believe that as early as 9,000 years ago, humans may have been deliberately breeding dogs to help them survive the hostile environment.

Based on reconstructions using fossil bones, including the 9,500-year-old remains of a dog eponymously named Zhokhov, scientists believe that some of the dogs many have resembled Siberian huskies; others may have been similar in size and shape to Alaskan malamutes. The smaller dogs would likely have been bred to pull sleds, while the larger dogs might have been bred to help hunt polar bears. In addition to canine bones, archaeologists found parts of sleds, including pieces of runners. Previously, the earliest identified example of selective breeding was herding dogs in the Near East about 7,000 years ago. Today's modern sled dog breeds—the Siberian husky, Alaskan malamute, and Greenland dog—share genetic heritage with the Zhohkov dogs.

Other interesting findings from Zhokhov: Sledge dogs have genetic adaptations to high-fat diets, like those found in polar bears and Arctic people. They also have genetic adaptations to physical exertion in cold temperatures. Researchers think that the insights gleaned from the canine fossil record on Zhokhov will help illuminate larger questions about where, when, and why dogs were domesticated. Most research has focused on Europe and East Asia, but the Arctic appears to be an interesting part of the puzzle.

See also Balto; Domestication; Landrace; Selective breeding; Tricolor coats

Zoomies

Zoomies, or "the zoomies," is the colloquial term used to describe a dog's sudden burst of high-energy, exuberant movement, often in a circular or back-and-forth pattern, and seemingly erupting out of nowhere. A dog might look like she has been stung in the behind by an invisible bee: she will launch forward, body hunched, tail tucked, ears pinned to the head. But instead of fear or pain, you will see on her face a distinct expression of joy as she careens around the room, weaving in and out of legs and furniture and other obstacles. Possession by the zoomies spirit usually only lasts for a couple of minutes, at most, and since zoomies amounts to an all-out sprint, a post-zoomies dog is often panting and tired. Dog owners report various triggers for zoomies: a dog being reunited with their human after an absence; a dog who

has just finished a bath; a dog who has gotten to choose which way to go at a trail junction; a dog who has just engaged in a bit of mischief.

The exact origin of the term "zoomies" is unknown, but it has been used in reference to pet animal behavior since the turn of the twenty-first century. (Cats engage in zoomies, too.) Because zoomies have not yet been carefully studied and analyzed by scientists, all we can do is speculate—and so much the better, because studying zoomies might just squeeze all the fun out of them.

No one knows for sure, but zoomies seem unmistakably like a bodily manifestation of canine happiness, a flow state of pure exuberant joy. Dog joy seems to be infectious. A human would be hard pressed to watch a zoomies episode without a huge grin spreading across the face and a distinct lifting of the spirit.

It is a happy coincidence that "Zoomies" is our final entry in *Dogpedia*. We can finish our excursion into the lives of these incredible creatures with a reminder that joy is an emotional experience shared by both dogs and humans, and joy is exquisitely infectious. Of the many, many gifts dogs give humans, joy is one of the most precious. Beyond the countless functional roles that dogs have filled during our thousands of years of collaboration, despite the daily practical help we two species offer each other, and despite the darker threads in our shared history, dogs give us a profound gift: joy. And joy is something we can and should offer them in return.

See also Joy; Play

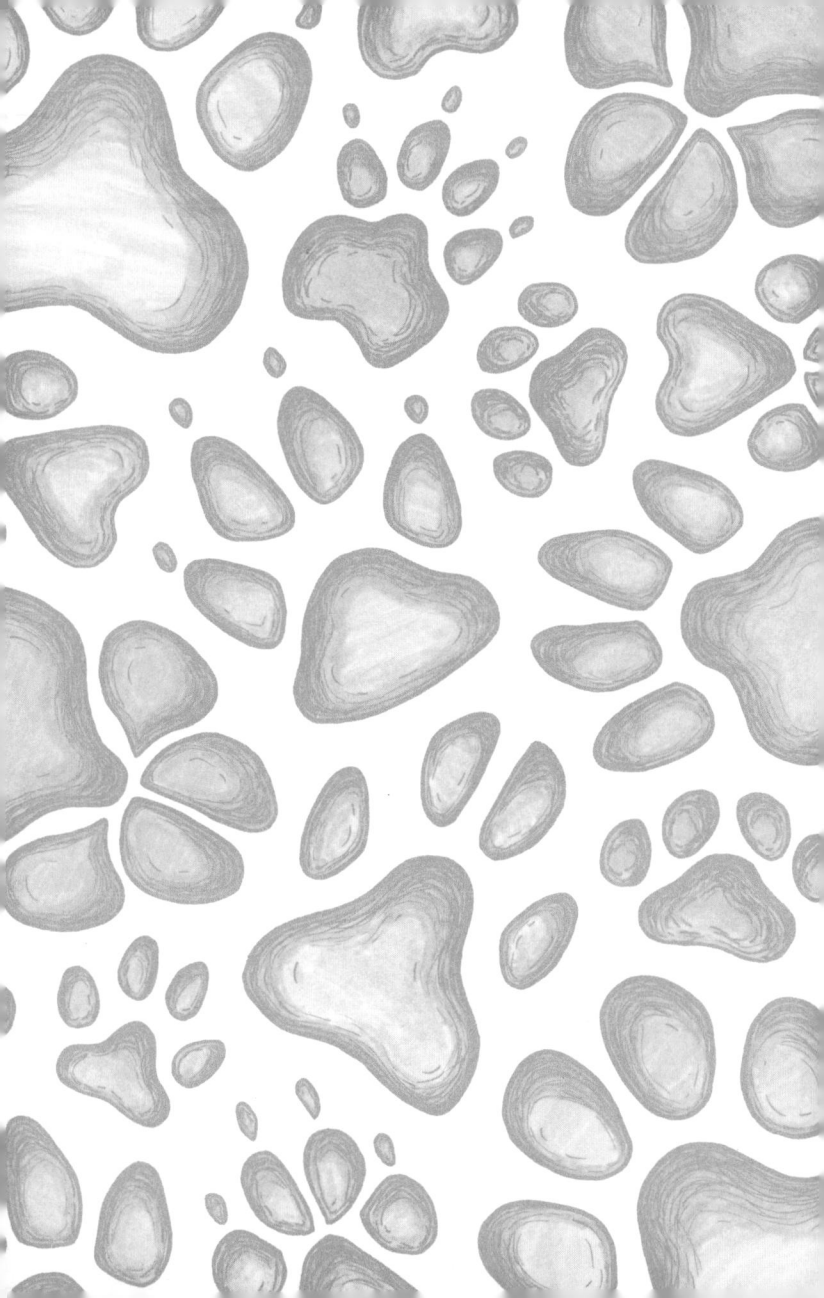

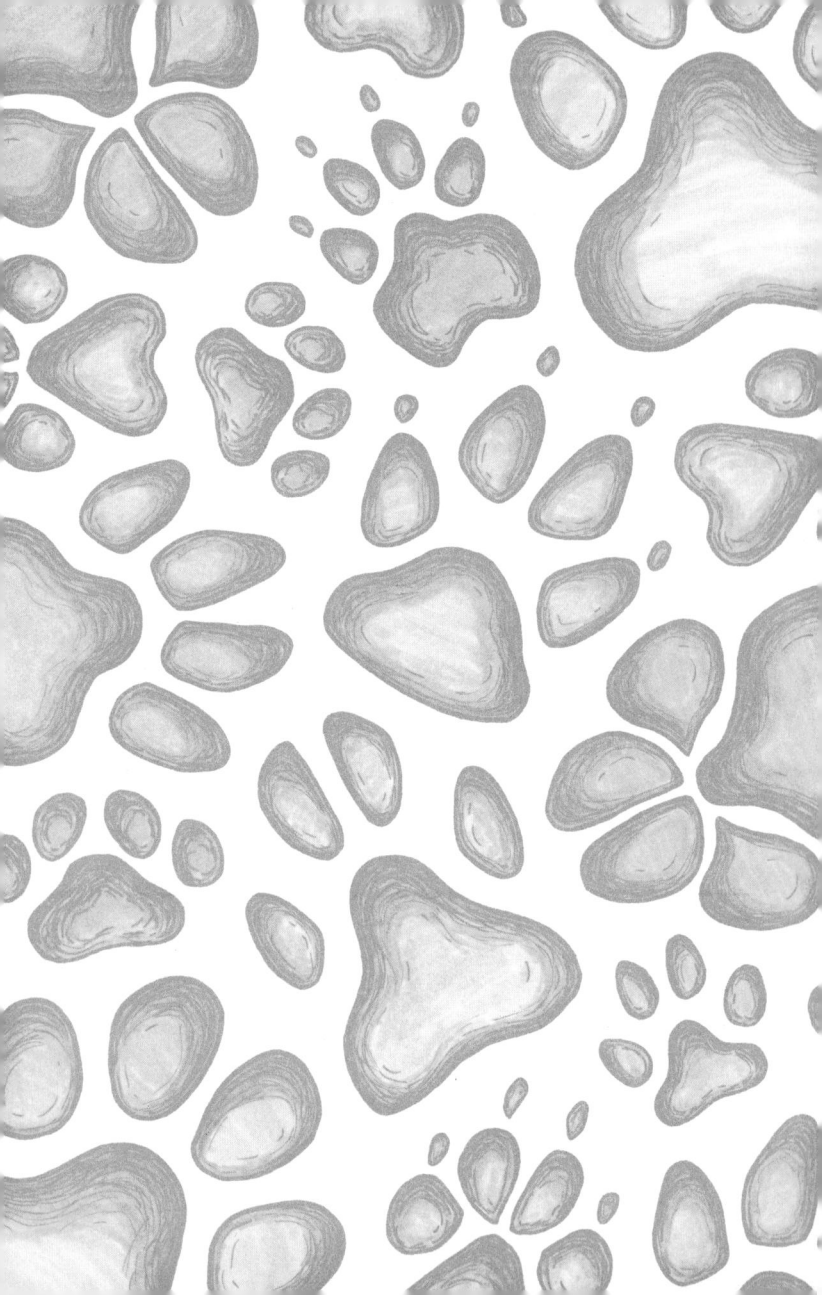